Chemical Applications
of Raman Spectroscopy

Chemical Applications
of Raman Spectroscopy

JEANETTE G. GRASSELLI and MARCIA K. SNAVELY

The Standard Oil Company
Research and Development Department
Cleveland, Ohio

BERNARD J. BULKIN

Chemistry Department
Polytechnic Institute of New York
Brooklyn, New York

A Wiley-Interscience Publication

JOHN WILEY & SONS, New York ● Chichester ● Brisbane ● Toronto

Library of Congress Cataloging in Publication Data:

Grasselli, Jeanette G.
   Chemical applications of Raman spectroscopy.

   "A Wiley-Interscience publication."
   Bibliography: p.
   Includes index.
   1. Raman spectroscopy.   I. Snavely, Marcia K.
II. Bulkin, Bernard J.   III. Title.

QD96.R34G7   543'.08584   81-1326
ISBN 0-471-08541-3   AACR2

Printed in the United States of America
10 9 8 7 6 5 4 3 2 1

Preface

In 1939 Hibben collected all the chemical applications of Raman spectroscopy to that date into a single volume, with a comprehensive list of all 1750 papers published on the subject. If anyone had attempted to update Hibben's work in 1959, it would have been more formidable, but still manageable. By 1979, when we were working on this book, we found relatively narrow areas of application on which more than 2000 papers had been published!

Raman spectroscopy has been periodically rejuvenated by advances in instrumentation that have permitted new applications. While the introduction of laser excitation is the best known of these, there have been numerous other instances. Holographic gratings freed the low frequency region from the plague of "ghosts," gallium arsenide photocathodes reopened red excitation, dye lasers advanced resonance Raman spectroscopy, and nonlinear techniques may yet be the means for overcoming fluorescence.

In this volume we have tried to say something about almost every application of Raman spectroscopy to chemistry. Should you find an application missing, we assure you that this results from our ignorance of it rather than from our judgment of its importance. By mentioning, even briefly, a very diverse set of applications, our goal is for this book to serve as an introduction to the literature. In this way we hope our work will be useful both to the practicing Raman spectroscopist confronted with a new area and to the chemist wondering whether Raman spectroscopy might help solve a problem.

At the same time, we treat a few topics in more depth to illustrate the various thought processes that go into interpreting Raman spectroscopic data. Naturally, for these topics we have chosen those subjects with which we are most familiar.

Several books have been devoted to aspects of the theory of the Raman effect. The brief introduction presented here should be sufficient for understanding the applications that follow. The discussion will be comprehensible to anyone who has completed a reasonably recent undergraduate physical chemistry course.

The future of Raman spectroscopic applications, now spanning biology, chemistry, and physics, is indeed a bright one. We prognosticate on this point throughout the book, especially in the closing chapter. We certainly hope to be able to check our sagacity in 1999.

JEANETTE G. GRASSELLI
MARCIA K. SNAVELY
BERNARD J. BULKIN

Brooklyn, New York
Cleveland, Ohio
June 1981

Contents

Chapter One

The Raman Effect

The Raman effect was discovered in India in 1928 by C. V. Raman, and, almost simultaneously, by G. Landsberg and L. Mandelstam in the Soviet Union. The attention paid to this discovery was very great, probably because of the intense interest in such inelastic phenomena generated by the earlier discovery of the Compton effect. Evidence of this attention is found in the early flurry of papers applying the Raman effect, technique development by such spectroscopic leaders as R. W. Wood, theoretical work on the effect in many laboratories, and the rapid award of the Nobel Prize to Raman in 1930.

In the first two decades after its discovery, Raman spectroscopy played a major role in many chemical structural studies. This is best appreciated by examination of the book by Hibben (1), *The Raman Effect and Its Chemical Applications,* which appeared in 1939 and contained 1753 references.

The next two decades, however, saw the rapid development of commercial infrared (IR) instrumentation, with almost no major advances in Raman instruments. The ease of obtaining vibrational spectroscopic data from IR quickly surpassed that of Raman spectroscopy when the Perkin-Elmer Model 21 was introduced.

The introduction of laser sources, beginning with the routine availability of 50 to 100 mW CW He—Ne lasers in the mid-1960s greatly increased the interest in Raman spectroscopy. Other advances in monochromators, detectors, and amplifiers contributed to the renewed interest, as did the routine availability of minicomputers. While IR spectrometers are still very much less expensive and more widely used, Raman spectroscopy fills a sizable role in vibrational spectroscopic work. In 1978–1979 about 3600 papers were published using Raman spectroscopy, compared with 8200 using IR spectroscopy.

1

1.1 COMPARING RAMAN AND INFRARED EXPERIMENTS

To better appreciate the reasons for this intense renewed interest in Raman spectroscopy, it is instructive to compare some of the advantages and disadvantages of Raman and IR spectroscopy with respect to instrumentation, sample handling, and applications. Tables 1.1 to 1.3, adapted from a comparison by Sloane (2), summarize such information. Since Raman scattering from molecular vibrations can be measured in the visible region of the spectrum, the optics of the instrument are relatively simple. Sensitive detectors with high signal-to-noise ratios are available. The intrinsic weakness of the Raman effect (Raman lines are about 10^{-6} the intensity of the exciting line) necessitates the use of an intense monochromatic light source, and, as such, the laser is ideal. A decided advantage of Raman spectroscopy is that the entire spectrum is obtained with the same instrument and cell, giving more information in a shorter time.

Infrared is applicable to almost any kind of sample but some materials (intractable polymers, single crystals, and aqueous solutions) are quite difficult to handle. With Raman spectroscopy, sample preparation is remarkably simple and

Table 1.1 Sample Handling

	Raman	Infrared
General applicability	95%	99%
Sample limitations	Color; fluorescence	Single crystals; metals; aqueous solutions
Ease of sample preparation	Very simple	Variable
Liquids	Very simple	Very simple
Powders	Very simple	More difficult
Single crystals	Very simple	Very difficult
Polymers	Very simple (but see sample limitations)	More difficult
Single fibers	Possible	Difficult
Gases and vapors	Now possible	Simple
Cells	Very simple (glass)	More complex (alkali halide)
Micro work	Good (< 1 μg)	Good (< 1 μg)
Trace work	Sometimes	Sometimes
High and low temperature	Moderately simple	Moderately simple

From ref. 2.

Table 1.2   Instrumentation

	Raman	Infrared
Relative complexity	Moderate	Slightly greater
Source	Laser	Blackbody or diode laser
Detector	Photomultiplier tube	Thermal, pyroelectric, bolometers
Resolution	ca. 0.25 cm^{-1}	ca. 0.05 cm^{-1}
Principal limitation	Energy	Energy
Wavenumber range	10–4000 Δ cm^{-1}	180–4000 cm^{-1} (one instrument) 10–400 cm^{-1} (second instrument or new beamsplitter, source and detector)
Purge requirement	No	Yes
Photometry	Scattering single beam	Absorption double beam

Adapted from ref. 2.

the capability for using glass or quartz cells is a marked advantage. Its principal limitation is with highly colored or fluorescing materials.

Both Raman and IR spectra are usually necessary to completely measure the vibrational modes of molecules. The techniques are complementary. Although some vibrational modes may be active in both the IR and Raman, these two forms of spectroscopy arise from different physical processes governed by specific selection rules, and the information content in the two techniques is a function of the molecular symmetry and polarity. Symmetric vibrations and

Table 1.3   Applications

	Raman	Infrared
Fingerprinting	Excellent	Excellent
Best vibrations	Symmetric	Asymmetric
Assignment work	Excellent	Very good
Group frequencies	Excellent	Excellent
Aqueous solutions	Very good	Very difficult
Quantitative analysis	More difficult	Good
Low frequency modes	Excellent	Difficult

From ref. 2.

nonpolar groups are most easily studied by Raman, antisymmetric vibrations and polar groups by infrared. At the empirical level, both techniques provide excellent "fingerprint" spectra for qualitative identification of molecules by the analytical chemist. Infrared spectroscopy holds an advantage in the huge number of reference spectra that are available, but group frequencies are equally well-recognized and useful in both methods. The possibility of examining aqueous solutions by Raman spectroscopy gives it a tremendous advantage over IR in biological and inorganic chemistry.

In this book, we highlight some aspects of modern Raman spectroscopy and its applications to chemical problems. It would be impossible to be comprehensive. Instead, we have provided a brief introduction to Raman spectroscopic instrumentation, techniques, and group frequencies, then selected a number of areas and examples of interest, in the hopes of giving a flavor of the type of work being done. The depth of treatment of particular areas is, to some extent, more a reflection of our own interests and expertise than of the regard in which we hold the work of our colleagues. For regular reviews of the state-of-the-art in Raman spectroscopy, the reader is referred to the April reviews in *Analytical Chemistry* (3).

1.2 BASIC PRINCIPLES OF RAMAN SPECTROSCOPY

The quantum mechanical theory of the Raman effect has been reviewed in some detail by Long (4). In this section, a brief introduction is given to the principle of Raman scattering. This should be helpful in putting Raman in perspective with relation to IR spectroscopy.

Infrared spectroscopy involves the actual energy levels falling in what is known as the IR region of the spectrum (10,000 to 10 cm^{-1}, with the region from 10,000 to 4000 known as the near infrared, 4000 to 400 the mid-infrared, or simply infrared, and 400 to 10 cm^{-1} the far infrared). Most such molecular energy levels are associated with internal vibrations of molecules. Rotations of molecules also occur in the lower wavenumber end, as do external molecular vibrations and librations in condensed phases. For a small number of molecules, there are electronic states in the IR region. Infrared spectroscopy is the measurement of the absorption or emission of photons to or from one of these states.

Raman spectroscopy is the inelastic scattering of photons. It does not involve absorption or emission from the energy levels directly, but rather implicates intermediate virtual states. The applications of Raman spectroscopy in this book are those in which the inelastic scattering transfers energy to or from the same energy levels described in the last paragraph for IR spectroscopy. However, the

Raman effect is much more general than this and can be observed with energy levels in any region of the spectrum.

A simplified energy level diagram for the Raman effect is shown in Fig. 1.1. This would be the diagram for a diatomic molecule in which only the $v = 1$ level is shown. If this level is of energy $h\nu_1$ above the ground state, then absorption of photons with frequency ν_1 could be observed in the IR spectrum (note that this absorption would be allowed only for a heteronuclear diatomic molecule). Suppose that radiation of energy $h\nu_0$ is incident on such a molecule. Many things can happen. The bulk of the radiation, for a nonabsorbing sample, will be transmitted. The radiation may also be absorbed, refracted, diffracted, or reflected. A small fraction of the radiation is elastically scattered, that is, it is reradiated in all directions unchanged in energy. This can be viewed for quantum mechanical purposes as the excitation of the molecule to a virtual state of energy $h\nu_0$ and the return to the ground state. This phenomenon is known as Rayleigh scattering.

A fraction of the radiation that is scattered will be inelastically scattered, giving rise to the Raman effect. Two distinct events are possible here, as shown in Fig. 1.1. Molecules in the ground state give rise to Raman scattering with energy $h(\nu_0 - \nu_1)$. Those in a vibrationally excited state can scatter inelastically back to the ground state, giving a Raman effect with energy $h(\nu_0 + \nu_1)$. These are known as Stokes and anti-Stokes Raman scattering, respectively. Together they comprise the Raman spectrum.

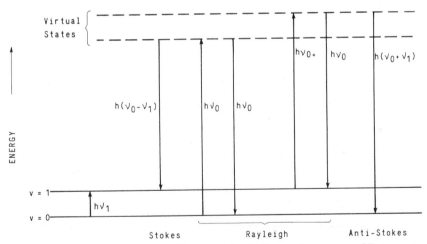

Figure 1.1 Energy level diagram illustrating the fundamental processes of Raman scattering. The exciting line is of energy h_ν. Raman bands appear at $h(\nu_0 - \nu_1)$ and $h(\nu_0 + \nu_1)$.

The ratio of Stokes to anti-Stokes intensity is governed by the temperature. At thermal equilibrium, this can be determined from the Boltzman distribution. When a system is not at thermal equilibrium, this ratio of intensities may be used to determine the deviation.

The result of making a measurement of the Raman spectrum of CCl_4 is shown in Fig. 1.2. This display is such that ν_0 is in the center of the figure. Stokes shifts are to the left, and anti-Stokes to the right. The anti-Stokes bands are clearly weaker than the Stokes, as expected. All of the shifts from ν_0 can be readily associated with the vibrational energy levels of the molecule.

The shifts in frequency observed in Raman spectroscopy do not always correspond to observable IR absorption bands. As previously noted, the two processes are governed by different selection rules relating to the interaction of the electrical nature of the normal vibration with the oscillating electric field of the electromagnetic radiation. Infrared absorption occurs when the radiation interacts with a normal mode of vibration in which the dipole moment of the molecule varies with time. Raman scattering occurs when the molecular motion produces a change in the polarizability of the molecule.

The fundamental differences in the processes that govern IR absorption and

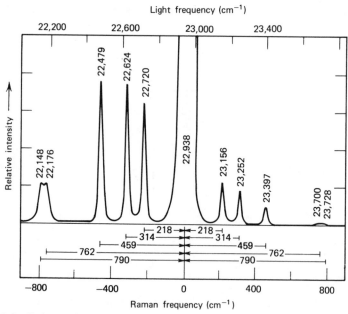

Figure 1.2 Stokes and anti-Stokes lines in the Raman spectrum of CCl_4—mercury arc excitation, 435.8 nm (22,938 cm^{-1}).

Raman scattering provide the basis for the appearance of bands or differences in band intensities in the spectra. This in turn facilitates the identification of various chemical groupings within the molecules. The development of group frequencies in the Raman has already been of considerable value in the utilization of this tool in analytical applications.

The vibrational analysis of molecules, through the use of group theory, allows the calculation of the number and activity of Raman and IR bands to be expected in the spectrum of a molecule. Some frequencies may appear in common, but as the molecular symmetry increases, the differences between the Raman and IR spectra also tend to increase, until mutual exclusion is attained for those molecules with a center of inversion. Thus the IR and Raman spectra provide a most important tool for determining molecular structure.

1.3   LIGHT SCATTERING AND THE RAMAN EFFECT

To qualitatively understand both the intensity and polarization of Raman bands, a simple introduction to light scattering is necessary. Consider a molecule to be a spherical, polarizable electron cloud, as in Fig. 1.3. When an electromagnetic wave of frequency v interacts with this cloud, it can cause it to oscillate at the same frequency. The extent to which this will occur depends on the polarizability or deformability of the electron cloud. An oscillating electron cloud radiates in all directions. This radiation is called light scattering.

Suppose that the incident beam is along the z axis, and is unpolarized, as

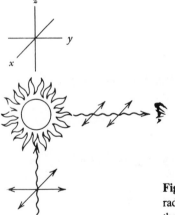

Figure 1.3   Scattering of unpolarized electromagnetic radiation, incident along the z direction, observed along the y direction. For a spherical scatterer, polarized scattered radiation is observed.

shown in Fig. 1.3. The incident beam has electric vectors along x and y, and so the electron cloud will also oscillate in the xy plane. An observer along the y direction can observe only electric vectors in the xz plane, since it is not possible to observe an electric vector along the direction of propagation. The result is that the observer along y would observe *polarized* (E_x) radiation resulting from the scattering of *unpolarized* light incident along z.

In the Raman effect, the incident beam couples to molecular vibrations (or other energy levels) to give inelastic scattering, that is, the scattered radiation is of a different energy from the incident beam. If the molecule is approximately spherical to begin with (e.g., CH_4) and the vibration is totally symmetric, then the polarization is maintained as before. However, if the vibration distorts the spherical symmetry, or if such symmetry does not exist to begin with, there may be significant intensity along the z direction in the observation along y.

Many texts have described the theory of Raman depolarization ratios (4–7). It suffices to summarize by saying that for (E_x) polarized incident radiation (the normal case for a modern laser source), incident along z, and analysis of the 90° scattered radiation along y, the depolarization ratio

$$\rho = \frac{I_z}{I_x}$$

ranges from 0 to 0.75. The range of values $0 < \rho < 0.75$ is obtained for vibrational modes belonging to the totally symmetric representation, while all other modes, referred to as depolarized, yield $\rho = 0.75$.

The Raman effect can be predicted by classical arguments. The electric field **E** associated with an incident beam of frequency ν_0 can be written in its time-dependent form as

$$\mathbf{E} = E_0 \cos 2\pi \, \nu_0 t$$

where E_0 is the amplitude of the wave.

When this oscillating field interacts with the polarizable electric field, it induces a dipole

$$\mathbf{M} = \alpha \mathbf{E}$$

where α is the polarizability of the material. The Raman effect results from the interaction of the polarizability with normal modes (Q) of vibration of the molecules. For the kth normal mode, we can write

$$\alpha = \alpha^0 + \frac{\partial \alpha}{\partial Q_k} Q_k + \cdots \text{ (higher order terms)}$$

where α^0 is the polarizability for the molecule with fixed nuclear positions. The normal modes must also be written in their time-dependent form as

$$Q_k = Q_k^0 \cos 2\pi \, \nu_k t$$

Substituting back into the expression for the induced dipole, we get

$$\mathbf{M} = \alpha^0 E_0 \cos 2\pi \, \nu_0 t + E_0 \left(\frac{\partial \alpha}{\partial Q_k} \right) Q_k^0 [\cos 2\pi(\nu_0 + \nu_k)t + \cos 2\pi(\nu_0 - \nu_k)t]$$

The first term is the unshifted (Rayleigh) light scattering, while the second term gives the Stokes and anti-Stokes Raman frequencies of $\nu_0 \pm \nu_k$. This classical argument fails to predict the difference in intensity between these two combinations, a difference that results from the Boltzman distribution of state populations.

In the general case, the induced moments should be written

$$\mathbf{M}_x = \alpha_{xx} \mathbf{E}_x + \alpha_{xy} \mathbf{E}_y + \alpha_{xz} \mathbf{E}_z$$

$$\mathbf{M}_y = \alpha_{yx} \mathbf{E}_x + \alpha_{yy} \mathbf{E}_y + \alpha_{yz} \mathbf{E}_z$$

$$\mathbf{M}_z = \alpha_{zx} \mathbf{E}_x + \alpha_{zy} \mathbf{E}_y + \alpha_{zz} \mathbf{E}_z$$

leading to an equation in matrix form:

$$\begin{pmatrix} M_x \\ M_y \\ M_z \end{pmatrix} = \begin{pmatrix} \alpha_{xx} & \alpha_{xy} & \alpha_{xz} \\ \alpha_{yx} & \alpha_{yy} & \alpha_{yz} \\ \alpha_{zx} & \alpha_{zy} & \alpha_{zz} \end{pmatrix} \begin{pmatrix} E_x \\ E_y \\ E_z \end{pmatrix}$$

In actual practice, the intensities measured in an experiment will depend on averages of the polarizability derivatives for an isotropic system such as a liquid or gas, or on a specific orientation for a system such as a single crystal.

1.4 SELECTION RULES

Group theoretical procedures allow one to derive selection rules for Raman scattering by molecules and crystals. Many texts give these procedures in detail

(4–7), and they need not be repeated here. It is useful, nonetheless, to indicate the connection of these rules to symmetry and quantum mechanics.

In any spectroscopy, selection rules arise from the integrals

$$\int \psi_0 \, \hat{O} \, \psi_e \, d\tau$$

where ψ_0 and ψ_e are the ground and excited state wave functions, and $\hat{O}$ is the operator involved in the spectroscopic transition. For this integral to be nonzero, that is, to be an allowed transition, it must be totally symmetric. This means that the product of three symmetries (ground state, excited state, operator) must be totally symmetric—invariant under all symmetry operations.

The origin of the difference between IR and Raman selection rules can now be explained. Infrared spectroscopic transitions originate from the dipole moment operator, with symmetry the same as vectors along x, y, and z. In the usual case, the ground state is totally symmetric; hence the product of the vibrational symmetry and dipole moment operator component must be totally symmetric. This is satisfied when they are the same, as the products x^2, y^2, z^2 will be symmetric. So the IR absorptions arise from vibrations with symmetry the same as x, y, z, and these are generally unsymmetrical vibrations.

In the Raman spectrum it is the polarizability that governs the symmetry. This has the same symmetry properties of the product of the Cartesian vectors x^2, y^2, z^2, xy, xz, yz. The Raman active vibrations will thus include those that are themselves totally symmetric, transforming as x^2, y^2, z^2. It should be noted that some vibrations may be inactive.

It is enlightening to scan the character tables of the point groups to see what generalizations concerning differences between IR and Raman selection rules emerge. Best known is that when there is a center of symmetry, vibrations cannot be active in both spectra. There are many other groups, *without a center of symmetry*, in which the selection rules are still distinct for IR and Raman spectra. Indeed mutual exclusion also occurs for such groups as C_{5h}, D_{5d}, D_{5h}, D_{6d}, and S_8. In other cases, some vibrations will appear in only one of the spectra. Examples of such groups are D_2, C_4, C_{3h}, D_3, D_{2d}, C_{4v}, D_4, D_{3h}, C_{6v}, T, T_d, O, and $C_{\infty v}$. In some of these cases (e.g., D_2, C_{3h}, D_3, D_{2d}) the distinction occurs in the totally symmetric representation not containing any component of the dipole moment operator. This facilitates vibrational assignments as these are also polarized vibrations. Another interesting case is C_1 symmetry, in which there is no symmetry, all vibrations are both IR and Raman active, and all Raman active vibrations are polarized. This becomes important when contrasted with other groups having a single symmetry element, C_S, C_2, and C_i. For C_S and C_2 all vibrations are again active, but some are depolarized in the Raman spectrum. For C_i mutual exclusion applies. Thus selection rules can be important in structure determination even for low symmetry molecules.

1.5 CRYSTALS

The use of Raman spectroscopy for the study of structure and lattice dynamics of single crystals was recognized very early, but the application to these problems only began to flourish in the laser era.

By appropriately selecting the orientation of the crystal, and the polarizations of exciting and scattered light, one can make a large number of independent observations on a single crystal. In contrast to liquids, where only averages of the polarizability tensor components are observable, in crystals each component may be isolated. Group theoretical methods can be used to predict the Raman activity of these components, and it is often possible to make predictions, for simple crystals, of the relative intensities as well. The procedures for doing this are given by Loudon (8). It may be useful to consult Wilkinson (9) for some worked examples.

For molecular crystals, three different symmetries may be important, namely, the unit cell (factor group) symmetry, the site symmetry of each molecule in the cell, and the molecular point group. Without knowledge of the potential energy, it is not possible to know in advance which of these will govern internal vibrations. External vibrations follow the factor group. To assist in deriving selection rules, correlation tables between the groups have been prepared (10). Their use makes the process a straightforward one.

REFERENCES

1. J. H. Hibben, *The Raman Effect and Its Chemical Applications*, Reinhold, New York, 1939.
2. H. Sloane, *Appl. Spectrosc.* **25,** 430 (1971).
3. D. J. Gardiner, *Anal. Chem.* **52,** 96R (1980).
4. D. A. Long, *Raman Spectroscopy*, McGraw-Hill, New York, 1977.
5. G. Herzberg, *Molecular Spectra and Molecular Structure*, Vol. 2, Van Nostrand, Princeton, 1945.
6. E. B. Wilson, J. C. Decius, and P. C. Cross, *Molecular Vibrations*, McGraw-Hill, New York, 1955.
7. S. Bhagavantam and T. Venkatarayudu, *Theory of Groups and Its Application to Physical Problems*, 3rd ed., Andhra University Press, Andhra, India, 1962 (reissued by Academic, New York, 1968).
8. R. Loudon, *Adv. Phys.* **13,** 423 (1964).
9. G. R. Wilkinson, *Molecular Spectroscopy*, Vol. 3, Chemical Society, London, 1973.
10. W. G. Fateley, N. T. McDevitt, and F. F. Bentley, *Appl. Spectrosc.* **25,** 155 (1971).

Chapter Two

Modern Raman Instruments and Techniques

This history of Raman spectroscopy has been greatly influenced by developments in the available instrumentation. Most aspects of this history are well-known, leading up to the production of commercial laser excited Raman spectrometers, with high quality double monochromators, photomultiplier tubes, and photon counting in the early 1960s (1).

We are now in the period of a second generation of laser excited spectrometers, offering far greater potential for observation of weak spectra (all Raman spectra seem to fall in this class) than previously. Some aspects of this development were expected and have gradually become widely available. Others appeared quite unexpectedly.

2.1 COMPONENTS OF MODERN RAMAN SPECTROMETERS

It was probably inevitable that the lower power helium-neon lasers, used in the early days of laser excited instruments, would give way to blue or green wavelength lasers. This has indeed been the case, with virtually all Raman instruments now using argon or krypton ion lasers. Light scattering increases as v^4, leading to a gain in Raman intensity. More sensitive detectors are also available.

Typical exciting wavelengths are shown in Table 2.1. Within the technology of Ar^+ and Kr^+ lasers, a recent development has been the availability of much higher powers, up to 15 W of CW power. It happens that very few condensed phase samples can withstand such power when it is focused to a diffraction

12

Table 2.1 Laser Emission Wavelengths Commonly Used to Excite Raman Spectra

Lasing Medium	Wavelengths (nm)
He—Ne	632.8
Ar$^+$	488.0
	514.5
Kr$^+$	530.9
	647.1

limited point, as is usual in Raman sampling. For certain cases, however, the high power is proving to be the difference between obtaining a spectrum and not obtaining one. This is particularly true for gas phase work. The higher power lasers are also useful for pumping dye lasers, so that the exciting frequency may be varied.

A potentially important Raman source is the frequency doubled synchronously pumped dye laser (2). This results from mode locking an argon ion laser, then going into a dye laser with a "cavity dumping" accessory. The result is tunable ultraviolet (UV) excitation of Raman spectra in the ultraviolet (285–308 nm). In addition, the short pulse widths (<10 ps) allow possible time resolved rejection of fluorescence.

A new photomultiplier tube is also part of this second generation instrument, featuring a Ga—As or multi-alkali photocathode surface. These tubes are typified by the RCA C31034 or the Hamamatsu 928. Two properties of these tubes make them desirable for Raman spectroscopy: they have a high absolute quantum efficiency, and the response is constant over the entire visible spectrum. These properties are illustrated in Fig. 2.1, where the RCA tube is compared to the ITT FW130 tube previously used in most instruments.

First generation laser excited instruments used gratings that were ruled. Now holographic gratings are available, and this means a far more perfect grating. Since stray light rejection and throughput are of critical importance to Raman spectroscopy, the holographic grating has meant a significant improvement in signal-to-noise ratios. It has also improved performance in the low frequency region, close to the Raman exciting line. Of interest is the production of concave holographic gratings, in addition to the conventional plane gratings. A monochromator that uses these gratings requires no additional optical elements other

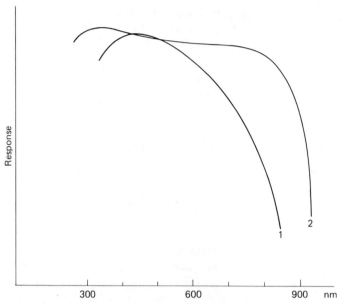

Figure 2.1 Photomultiplier response curves in the visible region for: (1) extended S-20 response, such as ITT FW-130; (2) Ga—As photocathode, such as RCA C31034.

than slits. This again reduces stray light. Such a monochromator is now commercially available (3). The stray light rejection in these monochromators is sufficiently good that Raman spectra can be obtained with a single monochromator, rather than the double or triple monochromators used previously. While most spectroscopists are still using the double monochromator, it seems likely that more will begin to take advantage of the single monochromator to achieve higher throughput, even at the expense of a modest increase in stray light.

In the earliest Raman instruments, DC amplification methods were used. Photon counting systems rapidly replaced these, and have become less expensive in recent years. Photon counters with the ability to count at very high rates are now routinely available. The digital signal of photon counting equipment readily interfaces to computers.

One result of using modern photon counters is that a very wide dynamic range of Raman scattering signals can be measured with good precision. This is important, as Raman scattering is observed over a very large range of intensities. It is probable that a comparable range exists in the infrared as well, but virtually all IR measurements are a ratio of two measurements. At high and low absorbances in the infrared, the errors in such measurements become extremely large.

As one might expect, Raman instruments are being routinely interfaced to minicomputers and to microprocessors. Several articles have discussed strategies and implementation methods (4). There are several reasons for doing this in Raman spectroscopy. Among these, signal averaging is foremost. Weak signals are readily enhanced by long integration times using the photon counting systems. The elimination of fluorescence by digital postprocessing of data is also possible (5). This is being accomplished by computing the Fourier transform of the Raman spectrum, picking the proper Fourier coefficients to filter the broad fluorescent background from the Raman signal, and recomputing the spectrum. Smoothing of data may also be accomplished in this way.

Polarization data, so important to applications of Raman spectroscopy, can also be readily extracted and displayed when data are available in digital form. One use of this is in separating the isotropic part of the polarizability, thus producing a display of only those modes belonging to the totally symmetric representation (6). In a conventional Raman measurement, two spectra, referred to as $I_\parallel$ (or I_{vv}) and $I_\perp$ (or I_{vh}), are usually obtained. A depolarization ratio,

$$\rho = \frac{3\beta^2}{45\alpha^2 + 4\beta^2} = \frac{I_\perp}{I_\parallel}$$

is then computed, where the α^2 terms are the isotropic portion of the polarizability, $\alpha = \frac{1}{3}(\alpha_1 + \alpha_2 + \alpha_3)$, and the β^2 terms are the anisotropic portion:

$$\beta^2 = \frac{1}{2}[(\alpha_1 - \alpha_2)^2 + (\alpha_2 - \alpha_3)^2 + (\alpha_3 - \alpha_1)^2]$$

In the technique referred to above, $I_\perp$ is scaled by 4/3 in the computer, then subtracted from $I_\parallel$ to produce a display proportional to $45\alpha^2$. Scherer (6) first applied this to vibrational assignments. Later, Bulkin et al. (7) pointed out that, for low symmetry molecules, structural questions often hinged on determining whether a molecule possessed C_s or C_1 symmetry. In the former case both polarized and depolarized bands should be observed; in the latter case only polarized. The polarizability separation technique is a convenient check on this. The relationship of polarization to structure is discussed at length in a later chapter.

Krishnan and Bulkin (8) have shown that, if the scaling factor is greater than 4/3, bands of any depolarization ratio may be removed from the spectrum, thus providing a means for effective resolution enhancement of data. They have also discussed some limitations of this technique.

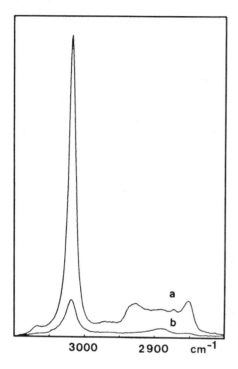

Figure 2.2 Raman spectrum of diplamitoyl lecithin in $CHCl_3$ solution at 3 cm^{-1} resolution: *(a)* $I_\parallel$ *(b)* $I_\perp$. (From ref. 8.)

Fig. 2.2 shows a complex envelope with many unresolved components. In each of the traces shown in Fig. 2.3 a different value of $1/\rho$ is used to scale $I_\perp$. As can be seen, each subtraction eliminates different bands from the band envelope.

Tobias et al. (9) have described a computer automated Raman spectrometer for obtaining difference spectra of materials as a means to increase sensitivity, particularly for the detection of minor reaction products in the presence of excess reactant. This technique has been applied principally to biological systems (10, 11).

Many other data processing applications have been accomplished with computer interfaced spectrometers. All of the work on Raman band shapes discussed later requires this. The Raman circular dichroism spectroscopy also requires digital data acquisition to be practical. Finally, to get good Raman intensity measurements, it is necessary to make a number of corrections to observed intensities. These have been discussed by Scherer and Kint (12). They are readily applied to an entire spectrum by digital postmultiplication.

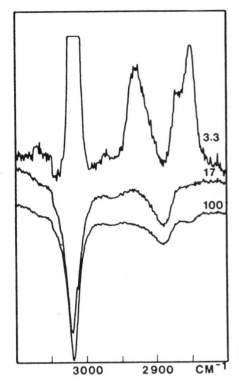

3000   2900   CM⁻¹

Figure 2.3 Spectra computed from $I_{\parallel}$ and $I_{\perp}$ spectra of Fig. 2.2, using $1/\rho$ values as indicated. (From ref. 8.)

2.2 TECHNIQUES

2.2.1 Sample Handling

At times the advantage of Raman spectroscopy over other analytical techniques lies in the ease of examining small, intractable, or difficult-to-handle samples in a relatively short time. Raman spectra can easily be run of gases, liquids, and solids. There has not been quite as much done with the gaseous state because it requires higher powered lasers and more complex sampling apparatus. However, special gas-handling cells and devices are available and have been described in the literature (13).

Liquids are extremely easy to examine by Raman spectroscopy. Glass sample bottles, flasks, and ampoules can be used directly if the glass itself does not contain impurities that cause fluorescence. The most common liquid devices, however, are 1.5 mm od capillary tubes in which samples are run neat or with solvents such as $CC\ell_4$ or CS_2. Water, an opaque solvent for IR work, is a poor scatterer and is used frequently for Raman spectral measurements. A rather flexible flowthrough cell has been described for observation of liquid samples without realignment (14). It was used to study electrode surfaces.

Solid state sampling is also straightforward in Raman spectroscopy. Samples can be tamped into an open-ended cavity for front surface illumination using a 180° mount platform or into a glass capillary tube for transverse excitation. Fibers, block specimens, and films can be studied directly without any special preparation. Minimum size of such samples is determined by the size of the focused laser spot and the difficulty of mounting the sample. Potassium bromide pellets mounted at 45° to the incident laser beam are also used quite frequently for Raman spectra of solid state samples.

2.2.2 Micro Work

The size of the laser beam allows the Raman spectra of small volumes of gases or liquids to be examined. Freeman and co-workers (15) have routinely obtained spectra from 2 nℓ liquid. Nyquist and Kagel (16) have achieved useful spectra from as little as 0.1 nℓ liquid in a 50 μm capillary. Rosasco and Simmons (17) have done experiments running Raman spectra of gases contained in glass bubbles where the effective scattering volumes were in the subnanoliter range. Barrett and Adams (18) have also reported the Raman spectra of O_2 and other gases from jets issuing from a nozzle in the 0.01 nℓ range.

2.2.3 The Raman Microprobe

One of the most interesting developments in the application of Raman spectroscopy to microchemical problems has been the coupling of a microscope to the Raman spectrometer (19–23). This allows one to examine a surface spectroscopically so as to discover nonuniformities that may be present and elucidate their chemical nature.

A schematic diagram of the commercially available Raman microprobe called MOLE (Molecular Optics Laser Examiner) is shown in Fig. 2.4. In this system, the laser beam can be scanned over the sample surface or positioned to any point on the surface. The beam area is approximately 1 μm^2.

Two modes of operation are possible. In the point mode, an area of interest on the surface is selected using the light microscope. The laser beam is positioned to this point and a spectrum is obtained.

18

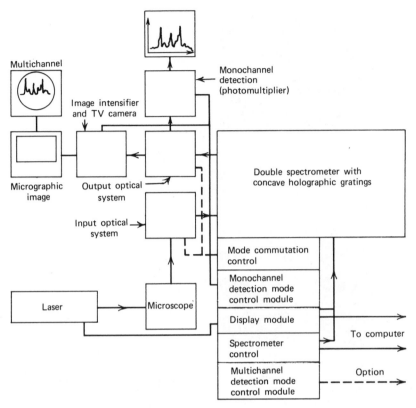

Figure 2.4 Schematic diagram of the Raman microprobe commercial instrument measured by Instruments SA. For discussion see text.

In the global mode, the spectrometer is set to a particular frequency characteristic of a species of interest and the beam is scanned over the surface. Using vidicon readout the two-dimensional spatial distribution of Raman scattering at the given wavenumber is determined.

One of the major applications of the Raman microprobe thus far has been to study inclusions in minerals. Particular inclusions can be isolated using the light microscope, then their spectra obtained in the point mode. Alternatively, in a sample with many diverse inclusions, the global mode can be used to construct a map of the surface.

Fig. 2.5 shows one example illustrating the modes of operation of the Raman microprobe. In the upper left frame, three particles are seen by the light microscope in the field of view. These give the composite Raman spectrum seen at

19

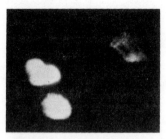 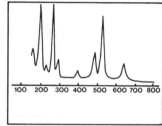

A sample is placed on the microscope slide and photographed (left). On the right is the total Raman spectrum obtained from the total field of view. Particles on the left are anatase (TiO_2); particle on the right is celestine ($SrSO_4$). Because the sum of the two spectra shown below is identical to the above total spectrum, there are no other chemical compounds present.

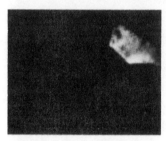 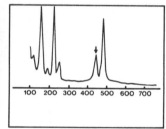

The laser light scattered from the particle on the right is analyzed for Raman spectrum to yield the partial spectrum shown at right. Image at left was obtained by setting the spectrometer for light at the frequency indicated by the arrow on the spectrum.

 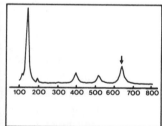

The laser light scattered from the two particles on the left is analyzed for the Raman spectrum shown at right. Image at left was obtained by setting spectrometer for light at the frequency indicated by the arrow on the spectrum.

Figure 2.5   Illustration of the use of the Raman microprobe.

the upper right. Individual particles can be isolated by the light microscope and their spectra obtained in the point mode. These spectra can be used to identify the particles as TiO_2 and $SrSO_4$. By setting the spectrometer in the global mode to the frequencies indicated by arrows in the figure, the images at center and lower left are obtained.

There are many types of samples to which the microprobe can be applied, including organic and inorganic species, polymers, salts of organic acids, and materials of biological importance such as urea and cholesterol (20). Vibrational spectra have also been obtained from individual microcrystals and fibers of sheet and chain silicate materials such as tremolite and talc (24, 25). Applications have included cement hardening heterogeneities, analysis of spots on thin layer plates, solid state reactions, analysis of microfossils, semiconductor device manufacturing control, and *in vivo* analysis of plant cells.

The Raman microscope has applicability in fields such as air pollution research, where the characterization of particulates is important. Fig. 2.6 shows the spectrum of an actual dust particle from urban air particulates. The particle was identified as $CaSO_4$ (anhydrite) from comparison of the major bands with a spectral reference. The broad bands at ~ 1400 and 1600 cm^{-1} have been attributed to degraded organic compounds or soot coating (26, 27).

The analytical potential of the Raman microprobe spectrometer in the trace characterization of polynuclear aromatic hydrocarbons appears promising (28). The PAHs representative of 3-, 4-, and 5-ring systems were examined as particles

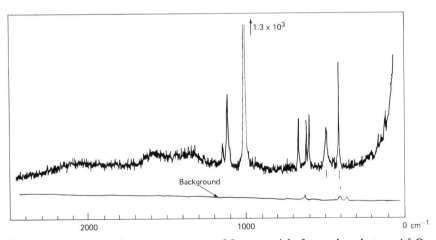

Figure 2.6   Raman microscope spectrum of 8 μm particle from urban dust on $A\ell_2O_3$ substrate. (From ref. 20.)

21

of 2 to 10 μm size. The detection limit was estimated to be from 10 to 100 pg for these environmentally significant hydrocarbons.

The Raman microprobe can be an important innovation in Raman spectroscopy even when the problem being studied does not involve microscopic examination. This is due to the optical design required in mating a microscope to a spectrometer. In this design the focusing and collection optics are optimized for a very small spatial region. This means that fluorescence or scattering from windows, substrates, and so on is collected much less efficiently than in the conventional Raman experiment. The spectra obtained with a diamond cell, for example (used to study Raman spectra at high pressures), are greatly improved in the Raman microscope because of the reduction of diamond fluorescence.

2.2.4 The Rotating Cell

Rotating cells were originally developed by Kiefer and Bernstein (29, 30) to study deeply colored, highly absorbing materials that otherwise would be destroyed by intense localized overheating caused by the laser. Cell designs have undergone many modifications (31–33), and their applications to problems involving liquids (29), solids (30, 34, 35), gases (36), and adsorbed species in catalysis (37) have grown tremendously.

The rotating cell is also a useful device for studying semimicro samples (38). Fig. 2.7 shows the Raman spectrum of 1.0 mg of $KMnO_4$ evaporated on a rotating sample holder from aqueous solution. Not only are the ν_1 and ν_3 bands at 840 cm^{-1} and 920 cm^{-1} observable (Fig. 2.7a), but so are overtones and combinations (Fig. 2.7b and c).

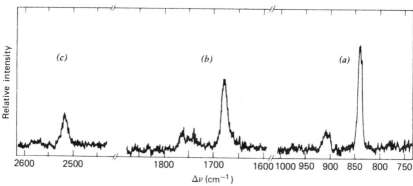

Figure 2.7 Raman spectrum of 1.0 mg of $KMnO_4$ evaporated on a rotating sample holder from aqueous solution. (From ref. 38.)

The rotating cell technique has also been adapted to record difference spectra of binary liquid systems using a divided cell and a gated electronic system (39). Unwanted solvent bands can thus be eliminated from the spectrum of a solution. Fig. 2.8 shows a 1 : 1 mixture of CCl_4 (A) and $CHCl_3$ (B) compared to the pure liquids. The upper field shows the Raman spectrum of the mixture ($A + B$); the second field shows the Raman spectrum of the reference liquid CCl_4

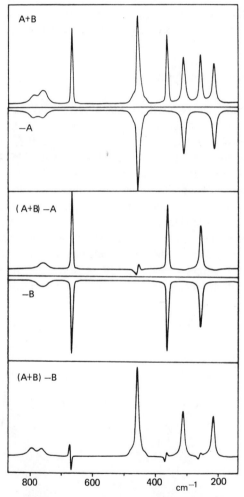

Figure 2.8 Raman and difference Raman spectra of CCl_4 *(A)*, $CHCl_3$ *(B)*, and a mixture of CCl_4 and $CHCl_3$ (*A* : *B* = 50 : 50 vol %); see text. (From ref. 39.)

($-A$), while the middle field displays the difference Raman spectrum of the mixture versus the reference liquid [$(A + B) - A$]. Spectrum $-B$ in Fig. 2.8 is the reference liquid CHCℓ_3, and the lower field shows the Raman difference spectrum of the mixture when CHCℓ_3 is the reference liquid. This difference Raman technique has also allowed accurate wavenumber shift measurements and corrections of the intensity error of Raman lines obtained from highly absorbing solutions (39).

Difference spectroscopy can be applied to solids as well as liquids by modification of a rotating solid sample cell (33). Bodenheimer et al. (40) have applied this technique to the study of single crystals under various orientations.

2.2.5 Quantitative Analysis

Raman spectroscopy can be used for quantitative work as well as qualitative studies. There are many different quantitative methods in use. Usually internal standards are added to unknowns, or bands are selected for reference that are unaffected by compositional changes (41–44), but Turner (45, 46) obtained good quantitative results with 10 mm fluorimeter cells in a cell replacement method.

Ratioing techniques have also been used quite successfully. Wancheck and Wolfram (47) reported a ratio method for determining the concentration of unreacted styrene monomer in latexes from an emulsion batch process for producing styrene/butadiene rubber. The latex samples were examined in melting point capillary tubes. Fig. 2.9 shows the C=C stretching region in the Raman spectrum for a standard latex sample containing 0.6 wt % residual styrene monomer, as measured by liquid chromatography and spiked with styrene to a total concen-

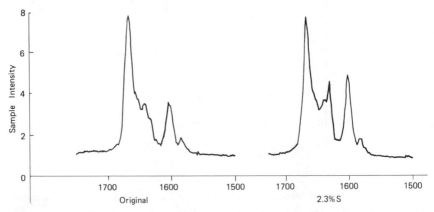

Figure 2.9 Addition of styrene to styrene/butadiene latexes used to measure free styrene concentration. (From ref. 47.)

tration of 2.3 wt %. The typical polybutadiene peaks at 1640 cm^{-1} (1,2-vinyl), 1652 (*cis*-1,4), and 1668 cm^{-1} (*trans*-1,4) are present plus the aromatic ring vibration of polystyrene at 1600 cm^{-1}. The styrene monomer peak occurs at 1632 cm^{-1}. As a general rule, in free radical polymerization of butadiene, the 1,4 mode of addition predominates, and the trans/cis ratio is determined by the polymerization temperature. For the commercial and experimental high butadiene latexes examined, there was little cis or 1,2-vinyl polymer, but more importantly, the distribution of unsaturated structures did not vary as determined by IR and nuclear magnetic resonance (NMR). Therefore, an analytical method was established that used the ratio of the 1632 cm^{-1} free styrene band to the 1668 cm^{-1} *trans*-1,4-polybutadiene band.

Calibration points were prepared by standard addition techniques of styrene monomer over the range 0.6 to 12.4 wt % styrene. The values of residual styrene obtained by the Raman technique were confirmed initially by liquid chromatography. The agreement between these methods was ±0.3%, and reproducibility on samples by the Raman method was ±0.1%.

The ability to examine water solutions directly was important in an industrial application where polymer plant personnel requested information on the extent of hydrolysis of acrylonitrile in a recycle monomer (48). Acrylonitrile is known to hydrolyze slowly in water, and to study this hydrolysis a capillary tube containing a 5% solution of acrylonitrile in water was monitored by Raman spectroscopy for a period of 29 days. Fig. 2.10 presents the data. The two hydrolysis products are hydracrylonitrile and acrylamide. Both of these could be followed quantitatively in samples from the plant with no more effort involved than filling a capillary tube and utilizing 20 min/day of Raman instrument time.

The band ratio technique has also been successfully employed to determine the composition of copolymers (5). Fig. 2.11 is the calibration curve for determining composition of acrylonitrile/styrene copolymers. In this case the Raman was calibrated against NMR and carbon, hydrogen, and nitrogen elemental analyses for a set of standards. Subsequent Raman results on actual samples could be obtained very rapidly.

Fig. 2.12 is another illustration of quantitative Raman spectroscopy. Without any special cells or techniques, benzene can be detected easily in carbon tetrachloride at concentration levels down to at least 110 ppm (49).

A sample of ^{13}C-enriched CHCℓ_3 is shown in Fig. 2.13, taken in the original manufacturer's ampoule (50). The symmetric C—Cl stretching vibration near 670 cm^{-1} is split into two components, which are assigned to molecules with ^{12}C and ^{13}C atoms, respectively. The relative peak heights of these two bands gives a ^{13}C content of 58.7%.

Quantitative Raman spectroscopy has also been applied to the determination of oxyanion impurities in reagent grade chemicals such as NaNO$_3$ (51). In this instance integrated band intensities as measured by a planimeter were used.

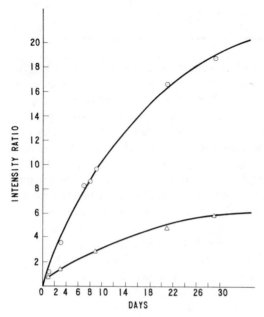

○ = INTENSITY 2230 CM⁻¹/INTENSITY 2220 CM⁻¹
(HO-CH₂-CH₂-CN)

△ = I₁₆₃₀ CM⁻¹/I₂₂₂₀ CM⁻¹

(H₂C=CH-C⟍O-NH₂)

Figure 2.10 Acrylonitrile hydrolysis study. The ratio of intensities yields the concentration of each product of the hydrolysis. (From ref. 48.)

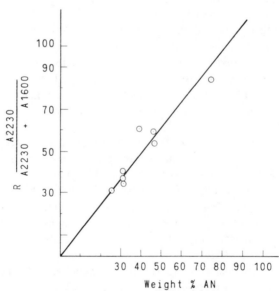

Figure 2.11 Calibration curve for acrylonitrile/styrene compositions. (From ref. 5.)

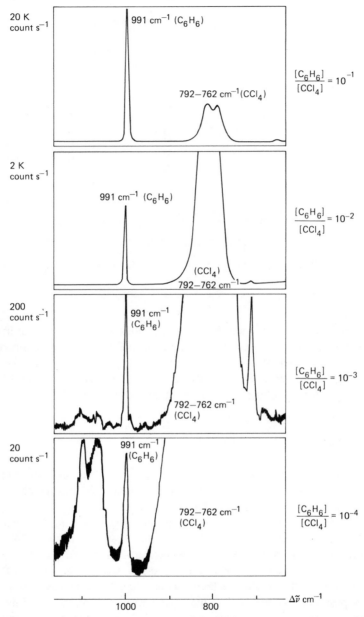

Figure 2.12 Quantitative analysis of benzene in carbon tetrachloride, 488 nm excitation. (From ref. 49.)

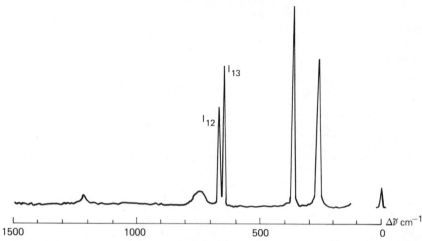

Figure 2.13   Raman spectrum of chloroform, 58.7% labeled with ^{13}C in the original ampoule. (From ref. 50.)

The rotating cell technique mentioned previously has also been utilized for quantitative work. A rotating cylindrical double cell with separate compartments for sample and reference obviates the need for an internal standard. This type of rotating cell has been used successfully with mixtures of carbon tetrachloride and toluene and on equilibrium studies of the dissociation of nitric acid (52).

2.2.6   Separated Fractions

Raman spectroscopy is an ideal technique to use in combination with other separation methods such as gas, thin-layer, or liquid chromatography because of the ability to look at very small sample volumes.

Since silica is a poor Raman scatterer, thin layer spots may be examined directly on a developed plate. Such a spectrum is shown in Fig. 2.14 (5). An evaporated gasoline sample was separated on a silica coated alumina strip and developed in 95% benzene/5% acetone. The spot was cut out, mounted on the 180° viewing platform, and the laser focused directly on the adsorbed layer by careful positioning of the microscope objective lens. Using the 514.5 nm argon ion line with 3 cm^{-1} slits, a spectrum easily identified as cresyl diphenyl phosphate was obtained. The ability to signal average spectra to improve sensitivity was also important. Huvenne et al. (53) have identified dodecane on silica gel plates by using a laser scanning technique in a backscattering geometry and data accumulation by a minicomputer.

Adams and Gardner (54) obtained the spectra of hexamethylenetetramine and several substituted benzophenones *in situ* on thin layer plates. They found that success depended upon the nature of the substrate (Kieselgel HR was better than Supreme, a grade of china clay, or Rutile), the eluant, the retention factor, fluorescence of both plate and sample, and the scattering efficiency of the sample itself. The Raman microprobe has also been used effectively for studying thin layer spots without further sample treatment.

A simple technique has been developed for examining liquid chromatography fractions (5). The collected peak is deposited stepwise under vacuum onto 15 mg of KBr. A 1.5 mm micro disc is prepared and the spectrum is obtained with 180° viewing. Fig. 2.15 shows the spectrum of 100 μg of a processing aid separated from a commercial polymer of liquid chromatography. With computer solvent subtraction and signal averaging, Raman spectroscopy has the potential to examine liquid chromatography fractions in glass vials with only minimal preconcentration. The direct coupling of an analytical liquid chromatograph to a Raman spectrometer has been described (55) using a flowthrough microcell.

Raman spectroscopy can also be used to characterize small quantities of sam-

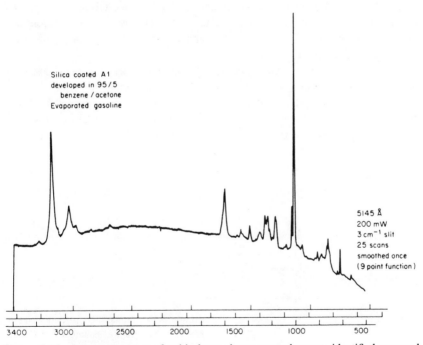

Silica coated A1
developed in 95/5
benzene / acetone
Evaporated gasoline

5145 Å
200 mW
3 cm⁻¹ slit
25 scans
smoothed once
(9 point function)

3400 3000 2500 2000 1500 1000 500

Figure 2.14 Raman spectrum of a thin layer chromatography spot, identified as cresyl diphenyl phosphate. (From ref. 5.)

29

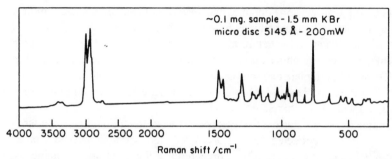

~0.1 mg. sample - 1.5 mm KBr
micro disc 5145 Å - 200mW

4000 3500 3000 2500 2000   1500   1000   500

Raman shift /cm⁻¹

Figure 2.15 Raman spectrum of a liquid chromatography fraction containing 100 μg of sample. (From ref. 5.)

ples trapped as gas chromatography effluents (56, 57). Fig. 2.16 shows a simple collection system consisting of a 0.3 mm id glass capillary inserted into a septum. Both ends of the capillary are open. When a peak appears at the detector, the capillary is held at the gas chromatography exit port with a cool moist tissue that condenses the liquid droplets along the walls. The walls can then be "swept" with a 0.1 mm id capillary to concentrate a slug of material at one end. This slug usually has a volume between 5 and 10 nℓ. The use of two different size capillaries permits a reasonable gas flow and yet allows small enough injections (2 to 5 μℓ) of sample to be practical.

A slightly different gas chromatography trapping system has been described by Nyquist and Kagel (58). In this system the sample is centrifuged into a constriction in the capillary after being collected in a portion cooled by liquid nitrogen. Fig. 2.17 shows two spectra obtained from the gas chromatography trapping of tribromopropane cuts. The upper spectrum is that of 2 μℓ of 1,2,2-

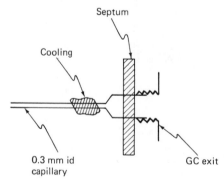

Septum

Cooling

0.3 mm id
capillary

GC exit

Figure 2.16 Apparatus for collecting gas chromatography effluents for Raman spectroscopy. (From ref. 56.)

30

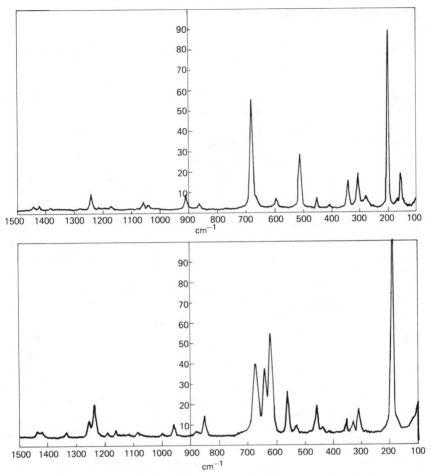

Figure 2.17 Raman spectra of trapped gas chromatography fractions: upper, 2 μl of 1,2,2-/-tribromopropane; lower, 20 nl of 1,2,3-tribromopropane. 632.8 nm He/Ne excitation. (From ref. 58.)

tribromopropane, while the lower represents 20 nℓ of the 1,2,3-isomer. The isomer was less than 1% of the original sample and was completely undetectable by IR spectroscopy.

The obvious advantage of employing any of the separatory techniques in obtaining Raman spectra is that the separated portion only contains one component and is relatively free from fluorescing impurities.

31

2.2.7 Resonance Raman

Resonance Raman scattering occurs when an electronic absorption band is located near the exciting line of the laser. When this happens, the form of the spectrum may change and the intensity increases greatly. A recent review by Morris and Wallan (59) outlines the theory of resonance Raman spectroscopy and describes many analytical applications. The use of resonance Raman spectroscopy in biochemistry and biology (60), for small molecules and ions (31), and for studies of the solid state (61) have been described.

Fig. 2.18 shows the resonance Raman spectrum of MnO_4^- in both H_2O and D_2O solution. The spectrum of this highly colored sample was taken in a rotating cell using 488 nm excitation (62). It is easy to see the very long overtone progressions of the 837.5 cm^{-1} MnO_4^- band in the spectrum, from which anharmonicity constants may be derived. Combining these results with extinction coefficients ε_0 for permanganate solutions at 488 nm, it was found that the intensity of the overtones varied as ε_0/V, where V is the vibrational quantum number. It is also interesting to note the differences in band widths, which may reflect changes in relaxation properties of the upper states (see Chapter 10).

The rotating cell has already been discussed, and is quite frequently used in resonance Raman spectroscopy. Shriver and Dunn (63) have pointed out the value of making measurements using a 180° scattering geometry rather than 90°. They show that in 180° backscattering work, the laser beam can be conveniently focused with a cylindrical lens, that is, focused to a line rather that the usual point focus. This disperses the energy over the medium, leading to reduced sample heating problems. In addition, it is a bit easier to correct intensities for sample absorption, or to minimize such absorption, in this geometry.

Another obvious use of resonance Raman spectroscopy is to obtain spectra of species at low concentrations enhanced in intensity. Studies of matrix-isolated species have been made (64). Resonance Raman scattering has been used extensively in biological studies for selective resonance enhancement in complex systems including whole bacteria (65), and thus has served as a subtle and sensitive probe in molecular structural and environmental studies. Examples are given in later chapters. The growing availability of tunable lasers will undoubtedly contribute to many more applications of this effect.

2.2.8 Image Intensifiers and Vidicons

Bridoux and Delhaye (66) have pioneered the use of multichannel detection systems for Raman spectroscopy. In a typical system, the scattered light, dispersed by the monochromator, is imaged on a multiple stage image intensifier

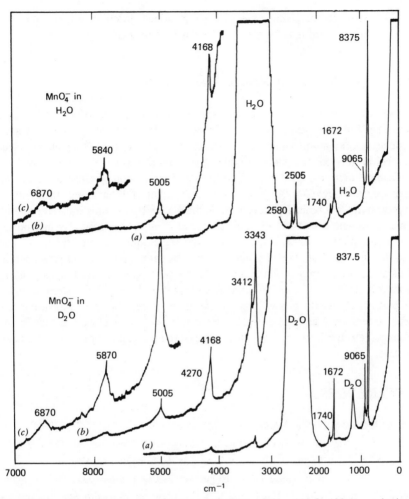

Figure 2.18 Resonance Raman spectrum of MnO_4^- in H_2O and D_2O, 488 nm excitation. (From ref. 62.)

tube and the resulting signal is focused on the photocathode of a Vidicon television tube.

A commercial detection system based on this principle, the OMA (Optical Multichannel Analyzer), is available from EG&G-PAR in Princeton, N.J.

There are many potential applications of multichannel techniques. With pulsed lasers, the OMA represents the best way of obtaining the Raman spectrum with

33

a single laser pulse. Using time gating techniques, in combination with short pulses, fluorescence may be separated from Raman scattering.

2.2.9 Rapid Scanning Raman Spectroscopy

Slower dynamic studies via Raman spectroscopy can be carried out by rapid scanning. For some years, Spex Industries has offered a rotating quartz refractor plate as an option with their double monochromator system. The plate is placed just before the exit slit and permits a rapid scan of a small spectral range.

Delhaye first described the possibilities and realization of rapid scanning Raman spectroscopy (67). Beny et al. (68) and Wallart (69) have described another approach to rapid scanning. It is used on the J-Y Optical double monochromator. This instrument operates with a cosecant bar cam arrangement. By interposing a second scanning motor, cam, and variable angle quoin between the lead screw and the cosecant bar, they are able to rock the gratings through a preset angle. This angle is determined by the quoin. In this way, spectra over a range varying from 5 to 1500 cm^{-1} can be obtained in less than 1 sec. The system seems to be particularly useful in studying the evolution of certain phase transitions that take place slowly and yield distinct Raman bands rather close together in frequency. Few applications of rapid scanning Raman spectroscopy have been made to date.

REFERENCES

1. B. J. Bulkin, *J. Chem. Educ.* **46,** A781, A859 (1969).
2. T. H. Bushaw, F. E. Lytle, and R. S. Tobias, *Appl. Spectrosc.* **32,** 585 (1978).
3. I. S. A. Jobin-Yvon, 16-18 rue du Canal, 91160 Longjumeau, France.
4. B. J. Bulkin, E. H. Cole, and A. Noguerola, *J. Chem. Educ.* **51,** A273 (1974).
5. J. G. Grasselli, M. A. Hazle, and L. E. Wolfram, in *Molecular Spectroscopy,* A. West, Ed., Heyden, New York, 1977.
6. J. R. Scherer, S. Kint, and G. F. Bailey, *J. Mol. Spectrosc.* **39,** 146 (1971).
7. B. J. Bulkin, D. L. Beveridge, and F. T. Prochaska, *J. Chem. Phys.* **55,** 5828 (1971).
8. K. Krishnan and B. J. Bulkin, *Appl. Spectrosc.* **32,** 338 (1978).
9. R. S. Tobias, T. H. Bushaw, and J. C. English, *Indian J. Pure Appl. Phys.* **16,** 401 (1978).
10. G. Y. H. Chu, S. Mansy, R. E. Duncan, and R. S. Tobias, *J. Amer. Chem. Soc.* **100,** 593 (1978).

11. S. Mansy, G. Y. H. Chu, R. E. Duncan, and R. S. Tobias, *J. Amer. Chem. Soc.* **100**, 607 (1978).

12. J. R. Scherer and S. Kint, *Appl. Opt.* **9**, 1615 (1970).

13. P. J. Hendra, in *Laboratory Methods in Infrared Spectroscopy*, R. E. Miller and B. C. Stace, Eds., Heyden, New York, 1972.

14. H. S. Gold, *Appl. Spectrosc.* **33**, 649 (1979).

15. S. K. Freeman, P. R. Reed, Jr., and D. O. Landon, *Mikrochim. Acta.* 288 (1972).

16. R. A. Nyquist and R. O. Kagel, in *Practical Spectroscopy*, Vol. 1, E. G. Brame and J. G. Grasselli, Eds., Dekker, New York, 1975.

17. G. J. Rosasco and H. H. Simmons, *Amer. Ceram. Soc. Bull.* **53**, 626 (1974).

18. J. J. Barrett and N. I. Adams, III, *J. Opt. Soc. Amer.* **58**, 3 (1968).

19. M. Delhaye and P. Dhamelincourt, *J. Raman Spectrosc.* **33**, 3 (1975).

20. G. J. Rosasco and E. S. Etz, *Res./Dev.* **28**, 20 (1977).

21. P. Dhamelincourt, F. Wallart, M. Le Clercq, A. Nguyen, and D. Landon, *Anal. Chem.* **51**, 414A (1979).

22. G. J. Rosasco, E. S. Etz, and W. Cassett, *Appl. Spectrosc.* **29**, 396 (1975).

23. B. W. Cook and J. D. Louden, *J. Raman Spectrosc.* **8**, 249 (1979).

24. J. J. Blaha and G. J. Rosasco, *Anal. Chem.* **50**, 892 (1978).

25. G. J. Rosasco and J. J. Blaha, *Appl. Spectrosc.* **34**, 140 (1980).

26. J. J. Blaha, G. J. Rosasco, and E. S. Etz, *Appl. Spectrosc.* **32**, 292 (1978).

27. E. S. Etz and G. J. Rosasco, *Environmental Analysis*, Academic, New York, 1977.

28. E. S. Etz, S. A. Wise, K. F. J. Heinrich, NBS Spec. Publ. 519 (Trace Org. Anal: New Front. Anal. Chem.), 723, 1979.

29. W. Kiefer and H. Bernstein, *Appl. Spectrosc.* **25**, 500 (1971).

30. W. Kiefer and H. Bernstein, *Appl. Spectrosc.* **25**, 609 (1971).

31. R. J. H. Clark, in *Advances in Infrared and Raman Spectroscopy*, Vol. 1, R. J. H. Clark and R. Hester, Eds. Heyden, London, 1975.

32. W. Kiefer, *Appl. Spectrosc.*, **28**, 115 (1974).

33. W. Kiefer, in *Advances in Infrared and Raman Spectroscopy*, Vol. 3, R. J. H. Clark and R. Hester, Eds., Heyden, London, 1977.

34. G. J. Sloane and R. Cook, *Appl. Spectrosc.* **26**, 589 (1972).

35. R. Carter and L. O'Hare, *Appl. Spectrosc.* **30**, 187 (1976).

36. R. Clark, O. Ellestad, and P. Mitchell, *Appl. Spectrosc.* **28**, 575 (1974).

37. C. P. Cheng, J. D. Ludowise, and G. L. Schrader, *Appl. Spectrosc.* **34**, 146 (1980).

38. G. J. Long, L. J. Basile, and J. R. Ferraro, *Appl. Spectrosc.*, **28**, 73 (1974).

39. W. Kiefer, *Appl. Spectrosc.*, **27**, 253 (1973).

40. J. Bodenheimer, B. Berenblut, and G. Wilkinson, *Chem. Phys. Lett.* **14**, 523 (1972).

41. D. E. Irish and H. Chen, *Appl. Spectrosc.* **25**, 1 (1971).

42. T. G. Chang and D. E. Irish, *J. Solution Chem.* **3**, 161 (1974).
43. A. K. Covington, M. L. Hassell, and D. Irish, *J. Solution Chem.* **3**, 629 (1974).
44. H. Baranska and A. Labudzinska, *Chem. Anal.* (Warsaw) **21**, 93 (1976).
45. D. Turner, *J. Chem. Soc., Faraday Trans. 2* **68**, 643 (1972).
46. D. Turner, *J. Chem. Soc., Faraday Trans. 1* **70**, 1346 (1974).
47. P. L. Wancheck and L. E. Wolfram, *Appl. Spectrosc.* **30**, 542 (1976).
48. J. G. Grasselli, M. A. S. Hazle, J. R. Mooney, and M. Mehicic, *Proc. 21st Colloq. Spectrosc. Int. and 6th Int. Conf. Atomic Spectrosc.*, Heyden, London, 1979.
49. D. Long, in *The Characterization of Chemical Purity, Organic Compounds*, L. Stavely, Ed., Butterworths, London, 1971.
50. B. Schrader, *Angew. Chem.* (Int. Ed. Engl.) **12**, 884 (1973).
51. D. Irish and J. Riddell, *Appl. Spectrosc.* **28**, 481 (1974).
52. A. K. Covington and J. Thain, *Appl. Spectrosc.* **29**, 386 (1975).
53. J. P. Huvenne, G. Vergoten, J. Charlier, Y. Moschetto, and G. Fleury, *C. R. Hebd. Seances Acad. Sci., Ser. C* **286** 633 (1978).
54. D. Adams and J. Gardner, *J. Chem. Soc., Perkin Trans. 2* **15**, 2278 (1972).
55. A. Chapput, B. Roussel, and J. Montastier, *C. R. Hebd. Seances Acad. Sci., Ser. C* **289***:11, 293 (1979)*.
56. B. J. Bulkin, K. Dill, and J. Dannenberg, *Anal. Chem.* **43**, 974 (1971).
57. R. Oertel and D. Myhre, *Anal. Chem.* **44**, 1589 (1972).
58. R. A. Nyquist and R. O. Kagel, in *Infrared and Raman Spectroscopy*, Vol. 1, Part B, E. G. Brame and J. G. Grasselli, Eds., Dekker, New York, 1977, p. 454.
59. M. D. Morris and D. J. Wallan, *Anal. Chem.* **51**, 182A (1979).
60. P. R. Carey, *Rev. Biophys.* **11**, 309 (1978).
61. A. Compaan, *Appl. Spectrosc. Rev.* **13**:2, 295 (1977).
62. W. Kiefer and H. J. Bernstein, *Chem. Phys. Lett.* **8**, 381 (1971).
63. D. F. Shriver and J. B. R. Dunn, *Appl. Spectrosc.* **28**, 319 (1974).
64. L. Andrews and R. C. Spiker, *J. Chem. Phys.* **59**, 1863 (1973).
65. W. F. Howard, Jr., W. H. Nelson, and J. F. Sperry, *Appl. Spectrosc.* **34**:1, 72 (1980).
66. M. Bridoux and M. Delhaye, *Nouv. Rev. Opt. Appl.* **1**, 23 (1970).
67. M. Delhaye, *Appl. Opt.* **7**, 2195 (1968).
68. J. M. Beny, B. Sombret, F. Wallart, and M. Leclercq, *J. Mol. Struct.* **45**, 349 (1978).
69. F. Wallart, Ph.D. Thesis, University of Lille, France, 1970.

Chapter Three

Applications to Organic Chemistry

Raman spectroscopy was widely applied to the solution of organic chemical problems during the 1930s. This has continued, with about 30% of all papers published on Raman spectroscopy each year being devoted to spectra of organics. The result is a substantial literature of group frequencies and applications, each area having been the subject of several complete books.

3.1 GROUP FREQUENCIES

In Raman as well as in IR spectroscopy, characteristic frequencies are found that are useful in chemical applications of these techniques. Raman scattering occurs if a bond is polarizable; IR absorption requires a change in dipole moment. For this reason, Raman is particularly informative about groups like —C—S—, —S—S—, —C—C—, —N=N—, and —C≡C—, whereas IR can be used to characterize groups like OH, C=O, P=O, S=O, and NO_2. In most cases the IR and Raman data are complementary, and it is helpful, even necessary, to have both for complete structural elucidation. In the past, IR group frequencies have been studied much more extensively (1–5), but many excellent texts and tables of Raman frequencies are now beginning to appear in the literature (6–9). Table 3.1 shows some characteristic bands that may be used for investigation using the Raman spectrum alone or a combination of Raman and IR spectra (6).

Raman spectroscopy is uniquely capable of characterizing many C=C stretching vibrations that generally occur near 1640 cm^{-1} and are often weak in the IR spectrum. In fact, when the band is symmetrically substituted, selection rules

Table 3.1 Characteristic Wavenumbers and Raman and Infrared Intensities of Groups in Organic Compounds

Vibration[a]	Region (cm^{-1})	Intensity[b] Raman	Intensity[b] Infrared
ν(O—H)	3650–3000	w	s
ν(N—H)	3500–3300	m	m
ν(≡C—H)	3300	w	s
ν(=C—H)	3100–3000	s	m
ν(—C—H)	3000–2800	s	s
ν(—S—H)	2600–2550	s	w
ν(C≡N)	2255–2220	m–s	s–0
ν(C≡C)	2250–2100	vs	w–0
ν(C=O)	1820–1680	s–w	vs
ν(C=C)	1900–1500	vs–m	0–w
ν(C=N)	1680–1610	s	m
ν(N=N), aliphatic substituent	1580–1550	m	0
ν(N=N), aromatic substituent	1440–1410	m	0
ν_a((C—)NO$_2$)	1590–1530	m	s
ν_s((C—)NO$_2$)	1380–1340	vs	m
ν_a((C—)SO$_2$(—C))	1350–1310	w–0	s
ν_s((C—)SO$_2$(—C))	1160–1120	s	s
ν((C—)SO(—C))	1070–1020	m	s
ν(C=S)	1250–1000	s	w
δ(CH$_2$), δ_a(CH$_3$)	1470–1400	m	m
δ_s(CH$_3$)	1380	m–w, s, if at C=C	s–m
ν(CC), aromatics	1600, 1580	s–m	m–s
	1500, 1450	m–w	m–s
	1000	s (in mono-; m-; 1,3,5-/- derivatives)	0–w
ν(CC), alicyclics, and aliphatic chains	1300–600	s–m	m–w
ν_a(C—O—C)	1150–1060	w	s
ν_s(C—O—C)	970–800	s–m	w–0
ν_a(Si—O—Si)	1110–1000	w–0	vs
ν_s(Si—O—Si)	550–450	vs	w–0
ν(O—O)	900–845	s	0–w
ν(S—S)	550–430	s	0–w
ν(Se—Se)	330–290	s	0–w

Table 3.1 (continued)

Vibration[a]	Region (cm^{-1})	Intensity[b]	
		Raman	Infrared
ν(C(aromatic)—S)	1100–1080	s	s–m
ν(C(aliphatic)—S)	790–630	s	s–m
ν(C—Cl)	800–550	s	s
ν(C—Br)	700–500	s	s
ν(C—I)	660–480	s	s
δ_s(CC), aliphatic chains			
C_n, $n = 3 \ldots 12$	400–250	s–m	w–0
$n > 12$	$2495/n$		
Lattice vibrations in molecular crystals (librations and translational vibrations)	200–20	vs–0	s–0

From ref. 6.

[a]ν stretching vibration, δ bending vibration, ν_s symmetric vibration, ν_a antisymmetric vibration.

[b]vs very strong, s strong, m medium, w weak, 0 very weak or inactive.

forbid any appearance of an IR band. It is this type of symmetric vibration with symmetric charge distributions that is very strong in the Raman spectrum.

Group frequency variations usually occur systematically, depending upon adjacent groups in a molecule. Thus a study of these variations is also a study of the adjacent groups. The dependence of the C≡C stretching frequency upon substituents is shown in Table 3.2.

Another area where Raman spectroscopy has a distinct advantage over IR is in the elucidation of the nature of sulfur bonding. SH, C—S, and S—S vibrations have been studied extensively (10) and are quite well-defined. The SH stretch at ~2580 cm^{-1} is weak in the IR spectrum but quite intense in the Raman, as shown in the spectra of allyl mercaptan, Fig. 3.1. Also intense in the Raman are C—S vibrations occurring between 570 and 785 cm^{-1}, which often appear as doublets or multiple bands due to the presence of rotational isomers. Organic sulfides exhibit both symmetric and antisymmetric C—S—C stretching vibrations in the region 570 to 800 cm^{-1}.

The antisymmetric C—S—C band is sometimes difficult to assign because it occurs in the same region as the CH$_2$ rocking vibration of aliphatic hydrocarbons and because there are sometimes multiple bands due to rotational isomers.

Organodisulfides generally have less complex spectra in the 700 to 800 cm^{-1} region than the corresponding organosulfides. There is no coupling between the

39

Table 3.2   Characteristic Wavenumbers in Raman Spectra From Carbon-Carbon Double Bond Stretching

General Ranges

H H C=C R H 1635–1650	R H C=C R H 1640–1660	H H C=C R R 1635–1660
R H C=C H R 1665–1680	R H C=C R R 1665–1695	R R C=C R R 1665–1685

Specific Cases (Substituted Ethylenes—CH_3, Halogen)[a]

	X = CH_3	X = $C\ell$	X = Br
H H C=C H X	1648	1601	1593
H X C=C H X	1658	1611	
H H C=C X X	1670	1590	1587
X H C=C H X	1684	1578	1582
X X C=C H X	1682	1582	
X X C=C X X	1676	1577	

Table 3.2 (continued)

Specific Cases (Substituted Ethylenes—Aromatic)[b]

ϕ H C=C H H 1634	ϕ H C=C H_3C H 1631	ϕ H C=C ϕ H 1610
ϕ H C=C H CH_3 1668	ϕ ϕ C=C H H 1629	ϕ H C=C H ϕ 1648

Adapted from data in ref. 8.
[a]All this group have depolarization ratios ≤ 0.1.
[b]All this group have depolarization ratios > 0.1.

in- and out-of phase C—S—S—C stretching vibrations, and only one band appears. The —S—S stretching band occurs between 500 and 550 cm^{-1}.

Table 3.3 summarizes the Raman frequencies of the C—S, S—S, and S—H vibrations. More detailed charts and discussions concerning the various sulfur-containing groups can be found in Nyquist and Kagel (10).

3.1.1 Carbon-Halogen Stretching

Carbon-halogen stretching bands occur between 1000 and 450 cm^{-1} in the Raman spectrum. In open chain, monohalogenated compounds, the stretching vibrations actually occur within the range 800 to 450 cm^{-1}. Once again, identification based on either the Raman or IR spectrum alone is unreliable. Many classes of compounds give strong IR bands (aromatics, alkynes, etc.) or strong Raman bands (branched alkanes, sulfur-containing groups) in this same region, whereas v(C—X) shows strong bands in both spectra. Evaluating both the IR and Raman spectra, it is possible to distinguish the halogen atoms and to detect the type of substitution present on the carbon atom attached to the halogen (11) (Fig. 3.2).

Comparison of corresponding C—F, C—Cℓ, and C—Br vibrations in similar compounds show that the intensity increases progressively from C—F toC—Br, which is exactly the reverse of the IR intensity behavior. This is predictable, though, because it becomes easier to distort the electron cloud about the C—X bond through the series C—F to C—I, while the dipole moment decreases through the same series.

41

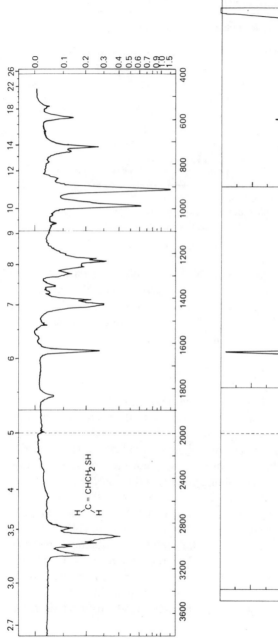

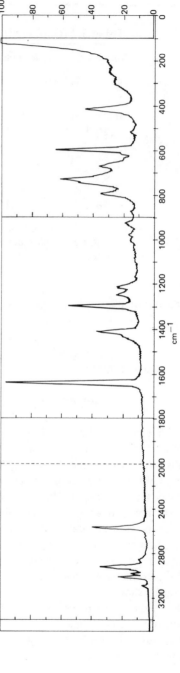

Figure 3.1 Allyl mercaptan: upper, IR spectrum, 10% in CCℓ₄ solution (3800 to 1333 cm⁻¹) and 10% in CS₂ solution (1333 to 4500 cm⁻¹), solvents are compensated; lower, Raman spectrum, neat liquid. (From ref. 10.)

Table 3.3 Raman SH, C—S, and S—S Vibrations

Organomercaptans (SH Stretching)

R—SH	2575–2584
HS—R—SH	2566–2582

Organomercaptans (C—S Stretching)

CH$_3$—S	705–785
RCH$_2$—S	660–670
R$_2$CH—S	600–630
R$_3$C—S	600–570
HS—R—SH	613–729

Organosulfides (C—S—C Stretching)

	Asym	Sym
R—S—R'	696–782	641–696

Organodisulfides (S—S Stretching)

R—S—S—R'	501–546

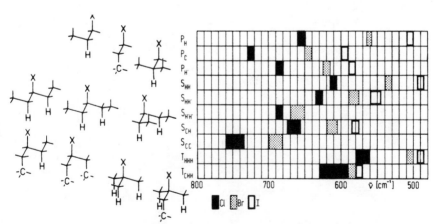

Figure 3.2 Correlation between carbon-halogen stretching frequency and the type of substitution present on the adjoining carbon atom. (From ref. 11.)

3.1.2 Triple Bond Stretching, C≡C and C≡N

The IR C≡C stretching vibration band near 2200 cm^{-1} is weak unless directly substituted with a polar group or atom, but the Raman band is nearly always intense. Likewise, the C≡N band is of variable intensity in the IR but is always strong in the Raman. In fact, if the C≡N group is α-substituted with a strong electronegative group such as chlorine, the IR shows little or no C≡N absorption near 2250 cm^{-1} (12). This is evident in the spectrum of α-chloroacetonitrile, shown in Fig. 3.3. However, the Raman spectrum, shown in the same figure,

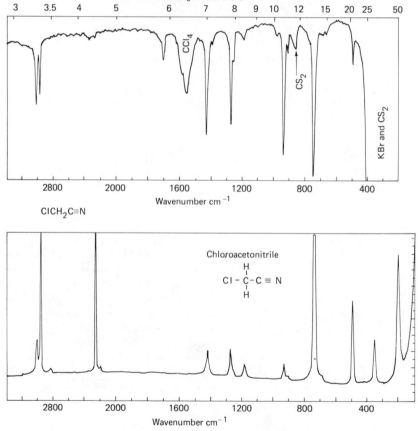

Figure 3.3 Infrared and Raman spectra of α-chloroacetonitrile. Due to α-carbon halogenation, intensity of —C≡N stretch is drastically reduced in the IR but retained in the Raman. (From ref. 12.)

has a very intense C≡N band. Thus it is more reliable to determine the presence or absence of the C≡C and C≡N groups with Raman than with IR spectroscopy.

3.1.3 Aromatic Structures

Raman spectroscopy is also extremely useful in the identification of aromatic structures. Infrared has classically been used for this purpose, because there are several regions of the spectrum with well-known characteristic absorptions. The CH stretching region above 3000 cm^{-1}, the overtone and combination bands in the 2000 to 1600 cm^{-1} region, the ring deformation bands in the 1500 and 1600 cm^{-1} regions, and the out-of-plane hydrogen deformation and ring puckering modes in the 900 to 700 cm^{-1} region indicate the type of substitution. Monosubstituted aromatic compounds typically show the latter absorptions at 750 cm^{-1} and 700 cm^{-1}, but the presence of electron-withdrawing groups on the ring can significantly disturb or alter these vibrations. In the Raman spectrum the characteristic ring modes are not affected by these substituents and it is easy to identify a monosubstituted benzene ring by the following bands:

CH stretch near	3060 cm^{-1}
Ring stretch doublet near	1600 cm^{-1}
Ring in-plane bending	618 cm^{-1}
In-plane CH deformation	1028 cm^{-1}
In-plane ring deformation	1000 cm^{-1}
	sh 995 cm^{-1}

Although not quite as definitive, other types of substituted benzenes also show well-resolved Raman bands. They are summarized in Fig. 3.4. Dollish et al. have discussed all of these bands more thoroughly (7).

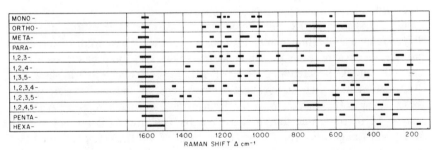

Figure 3.4 Characteristic frequencies in the Raman spectra of substituted benzenes. (From ref. 7.)

3.1.4 Other Characteristic Vibrations

Some groups can *only* be identified by Raman spectroscopy. For example, there are no characteristic IR group frequencies for Se—Se, S—S, and O—O. The O—O stretch, which gives rise to a strong band in the Raman spectrum near 700 to 900 cm^{-1}, can be used for the identification of peroxides, peracids, and peresters (6, 13).

Other groups such as C=O, N=N, C≡N, NO_2, NH, and OH manifest themselves in both the Raman and IR spectra. The presence or absence of a band, as well as its relative intensity, aids in structure elucidation. In addition to the carbon-halogen examples already mentioned, the intensities in the Raman spectrum of bands arising from functional groups like X—H, OH, and C—H are often the opposite of the IR. Several authors (6, 9, 12) have shown how Raman and IR spectroscopic data may be integrated in structure determinations.

3.2 PROBLEMS OF STRUCTURE AND CONFORMATION

3.2.1 Structure Elucidation

Both the IR and Raman spectra of a molecule are widely used in organic chemistry for structure elucidation. The differences in relative intensities of the various functional group peaks between the IR and Raman spectra have already been noted. It is often observed that the more unsymmetrically substituted a given bond, the greater the IR intensity and the weaker the Raman spectrum. Conversely, the more symmetrically substituted the groups, the stronger the Raman scattering.

Symmetry selectivity generally results in a simpler Raman spectrum than IR, a factor that can be extremely useful in structure elucidation. Many examples of this natural partnership for general vibrational analysis have been given by Washburn (14).

Fig. 3.5 shows the IR and Raman spectra of crystalline cystine (6). The NH_3^+ stretching vibration completely dominates the IR spectrum in the 3000 cm^{-1} region, whereas the Raman spectrum shows two sharp bands associated with CH and CH_2 stretching. Both the IR and Raman spectra show the NH_3^+ deformation and antisymmetric vibrations of the carboxylate group —CO_2^- near 1600 cm^{-1}, but in the Raman they are much weaker. A strong band at 1410 cm^{-1}, due to symmetric carboxylate stretch, is present in both the IR and Raman spectra. The strongest band of the Raman spectrum, however, occurs at 410 cm^{-1}, due to —S—S— stretch; this band is not nearly as apparent in the IR. Figure 3.5 also illustrates the ability to observe low frequency modes readily in the Raman spectrum.

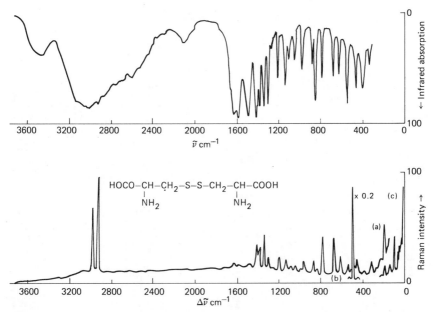

Figure 3.5 Infrared and Raman spectra of crystalline cystine. Infrared spectrum of 0.8 mg in 400 mg KI. Raman spectrum: *(a)* 2 mg, amplification × 1; *(b)* 2 mg, amplification × 0.2; *(c)* disc, 140 mg, amplification × 0.3. (From ref. 6.)

Fig. 3.6 shows the IR and Raman spectra of 5,5-/-dimethyl-3-phenylpyran-2(5H)-one. The greater simplicity of the Raman spectrum is quite evident. The C=C stretching vibration at 1640 cm^{-1} is very strong in the Raman but barely visible in the IR spectrum. On the other hand, the C=O stretch at 1700 cm^{-1} is only observed in the IR spectrum. The IR bands at 750 and 700 cm^{-1}, due to a monosubstituted benzene ring, are present but difficult to identify in such a complex spectrum. The Raman spectrum more clearly defines the substitution pattern from the ring doublet near 1600 cm^{-1}, the ring in-plane bending near 618 cm^{-1}, and ring deformation at 1000 cm^{-1} with a shoulder at 995 cm^{-1}.

3.2.2 Studies of Structurally Related Compounds

There have also been extensive studies of the Raman effect for structurally related compounds. Alkyl benzenes with 3 to 15 carbon atoms in the alkyl chain were investigated by Behroozi et al. (15) The chloromethyl group attached to an aliphatic hydrocarbon chain was examined by Wang and Mannion (16). Lere-

47

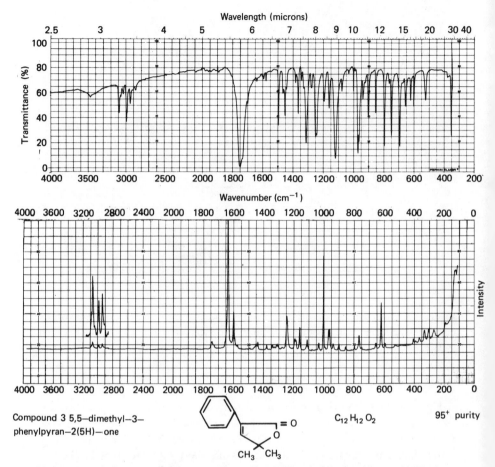

Figure 3.6 Infrared and Raman spectra of 5,5-dimethyl-3-phenylpyran-2(5H)-one. (From ref. 14.)

Porte et al. studied the CH_2 vibration of ethane disubstituted by polar groups such as OH and halogens (17). Nyquist (18) established the characteristic Raman bands for phthalate esters from a study of 21 compounds.

Freeman and Mayo (19) examined the thiomethyl group in various monosulfides, disulfides, and trisulfides. They also reported on the spectra of 70 di- and trisubstituted acyclic and cyclic compounds containing methyl groups situated on ethylenic carbon atoms (20). A good correlation was obtained between the number of methyls directly attached to ethylenic carbon atoms and the intensity ratio of the methyl symmetric deformation mode (ca. 1375 cm^{-1}) to a band at 1440 cm^{-1}.

Saturated long chain fatty acids have been studied by Warren and Hooper (21) for a relationship between wavenumber shifts of the longitudinal acoustic vibrations of the carbon skeleton (called accordion modes) and the chain length (C_{12}-C_{24}) of the acids.

There have been a number of studies of benzene derivatives that provide information on both the nature of the substitution pattern and the interactions of the substituents (10, 22). In these cases, however, one band alone is of little value—the existence, position, and intensity of a number of bands have to be considered. This type of information can be classified into simple yes and no answers that provide a scheme for the determination of the substitution type (22), as shown in Fig. 3.7. Such flow diagrams have also been devised by Willis et al. for barbiturates (23) and Oertel and Myhre for substituted pyrazines (24).

Vissir and Van der Maas have published a series of papers describing systematic procedures for the interpretation of Raman spectra of carbon-hydrogen and alcoholic compounds (25), ethers (based on 80 compounds) (26), carbonyl-containing compounds (based on 220 ketones, aldehydes, esters, and acids) (27), and nitrogen-containing organic compounds (based on 79 amines, pyridines, cyanides, amides, and nitro compounds) (28). A total of 606 compounds, including all groups, were examined. A Fortran program is available that contains 376 question or Q elements and 178 information or I elements. The authors plan to extend the system to other functionalities, setting up rules and procedures to overcome interferences resulting from the larger number of functionalities and related frequency intervals.

A similar systematic empirical evaluation of frequencies and intensities has been developed for Raman steroid spectra (29–31). It enables predictions about the number and intensities of multiple bonds and about the cis and trans coupling of rings *A* and *B* by using the CH_2 and CH_3 vibrations near 1450 cm^{-1} as reference groups. If these are the strongest bands in the spectrum, a saturated structure is indicated. If there are only bands stronger than the reference group above 1450 cm^{-1}, isolated or conjugated double bonds at the five-membered ring *D* or homoannular dienes are likely. The presence of conjugated double bonds in or at the six-membered rings leads to stronger bands both above and below 1450 cm^{-1}. Information concerning the nature and position of these structural elements can be obtained from the number and frequencies of these bands (6).

All of these logical analytical schemes can readily be adapted to a computer and, together with IR, UV, NMR, or mass spectral data, can be used for the identification of unknown compounds.

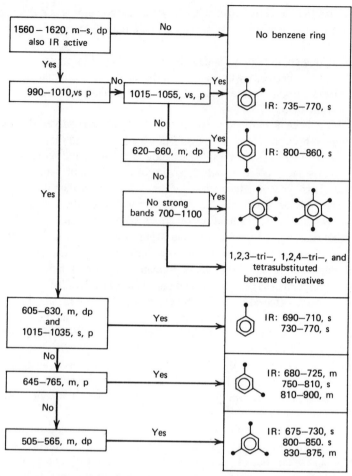

Figure 3.7 Scheme for the determination of the substitution pattern of benzene derivatives with the aid of characteristic bands in the Raman spectrum. Intensities: m, medium; s, strong; vs, very strong. Degree of polarization: p, polarized; dp, depolarized. IR denotes characteristic band in the IR spectrum. (From ref. 22.)

3.2.3 Conformation Studies and Molecular Symmetry

Raman spectral data utilizing molecular symmetry and normal vibration frequencies of representative organic molecules can be an aid to conformation studies and subsequent predictions of the expected spectra of larger and more complex molecules. The group theoretical method for calculating the number of vibrational modes in each symmetry class and the number of corresponding lines in the spectrum has been explained by a number of authors including Suschchinskii (32) and Cotton (33).

Many papers have been published involving conformational studies; only a few examples are listed here. Buge et al. (34) studied the IR and Raman spectra of liquid and solid diethylmaleate and fumarate. They confirmed a previously reported conformational equilibrium for the diethylfumarate and found at least two conformers in liquid dialkylmaleates.

Manley and Martin (35) recorded the IR and Raman polarized spectra of methyl, *n*-butyl (BMA), and *n*-octyl (OMA) methacrylates. They proposed complete assignments on the basis of C_S symmetry. They also discussed assignment of the torsional skeletal modes and interpretation of additional bands in the spectra of BMA and OMA.

Maillois et al. (36) investigated the Raman spectra of aqueous and D_2O solutions of fumaric and maleic acids. When combined with IR data, the results indicated that the fumaric acid was planar (C_{2h} symmetry) and that the maleic acid was nonplanar, belonging to either the C_S or C_2 group.

Lewis and Laane (37) have recorded the vibrational spectra of 3-cyclopentene-1-one in the vapor, liquid, and solid phases and carried out a complete vibrational analysis of the molecule. The spectra are in agreement with the planar C_{2v} structure. Many more examples of conformational studies of this sort are given in the annual reviews of Raman literature published by *Analytical Chemistry* (38–41).

Another area where Raman spectroscopy can be applied is the detailed study

Table 3.4 Potential Constants and Most Stable Forms

Molecule	Potential Constant (cm^{-1})					Most Stable Form
	V_1	V_2	V_3	V_4	V_6	
1,3-Butadiene	600	2068	278	-49		Trans
Glyoxal	1182 ± 10	1114 ± 10	0	-56 ± 4	0	Trans
Acrolein	306	1919	338	-96	-57	Trans
Styrene	0	623 ± 8	0	27 ± 3	0	Planar

Molecule	Potential Constant (cm^{-1})				Most Stable Form
	V_1	V_2	V_3	V_6	
$CH_3CH_2NH_2$	218 ± 52	0	251 ± 17	-52 ± 11	Trans
$CH_3CH_2PH_2$	0	270 ± 27	830 ± 17	-58 ± 10	Trans
CH_3CH_2SH		171 ± 2	484 ± 1	-21 ± 1	Gauche
CH_3CH_2SeH		-96 ± 1	432 ± 1	-20 ± 2	Gauche

51

of low frequency intramolecular modes, the knowledge of which is essential to a complete understanding of the properties and behavior of organic compounds. Molecular potential functions, which are related to energy barriers of various conformational states, can be obtained from the overtones of ring-puckering and internal torsional modes, and thus the conformation of the lowest energy state can be determined. Studies have been made on molecules such as 1,3-butadiene (42), styrene (43), glyoxal (44), acrolein (45), and a series of substituted ethanes (46–48). Potential constants (cm^{-1}) and the most stable forms for the molecules studied are tabulated in Table 3.4. Ring-puckering vibrations have been identified and inversion barriers calculated also from the spectra of various small ring compounds (49–51).

3.3 PHYSICAL PROPERTIES

Raman spectroscopy can also be used to study various types of physical properties—hydrogen bond strengths, acid dissociation constants, phase transitions, and energy differences between rotational isomers.

Perchard and Perchard (52) studied several OH (OD) stretching vibrations of alcohols for frequency shifts as a function of temperature and intensity and for polarization changes as a function of temperature and physical state. Alcohols were also examined for hydrogen bonding effects by Lavrik and Naberulchin (53), and hydrogen bonding in aqueous acid solutions was discussed by Pernall et al. (54).

Halogenated cyclohexanes have been studied in the vapor phase, in melts, in solutions, and in amorphous solids for conformational information (48, 49). This work was later extended to include cyano and isocyanatocyclohexane (55, 56).

The energy differences between rotational isomers of 2-chloroethanol and 2-bromoethanol (57) and ethylene glycol (58) have been determined from their Raman spectra.

Variable temperature IR and Raman studies have enabled complete vibrational assignments to be made and thermodynamic functions to be calculated for buta-1,3-diene and 2-methylbuta-1-1,3-diene (isoprene) (59). These thermodynamic functions, in turn, have permitted the equilibrium ratios between the *s*-trans and *s*-cis conformers at normal temperatures to be established.

3.4 UNSTABLE SPECIES: CHEMICAL AND PHYSICAL

Raman spectroscopy is becoming an important technique for the study of short lived species. These can be excited states or reaction intermediates.

In each case the strategy is to excite the molecules and probe their Raman spectra using two different energy pulses of the same short pulse laser, detecting with a multichannel detector such as a vidicon (cf. Chapter 2). The key to obtaining spectra of a small quantity of the unstable species in the presence of the stable species is to be in resonance with an electronic state of the desired moiety.

This is perhaps most obvious in the case of free radicals. Wilbrandt et al. (60, 61) have shown that radicals containing S—S linkages can easily be detected in resonance. As noted elsewhere in this chapter, this group has a high Raman cross section, and its stretching frequency is very sensitive to structure. In Wilbrandt's system the chemistry is carried out by electron pulses (62).

Mayer et al. (63) have studied many radical ions. They have been able to make careful vibrational assignments and deduce structural features.

Atkinson and co-workers (64) have also studied organic intermediates formed by pulsed radiolysis of a sample, concentrating their efforts on obtaining series of time resolved spectra. This work has been extended (65) to the study of excited states of aromatic species, such as chrysene, in the microsecond time range. It is possible that the sensitivity can be further enhanced by use of nonlinear techniques (66).

Yet another approach to the study of unstable species is matrix isolation. Andrews and co-workers (67, 68) have studied a number of species, such as alkali metal halides $M^+X_2^-$, again taking advantage of resonance Raman excitation. They have also studied rare gas halides (69) in this way.

Campion et al. (70) have used resonance Raman spectroscopy with the OMA to study kinetics of reactions connected with the vision process. In their work a CW laser is used. However, by interposing a rotating disk with a small slit in it, a pulse of several microseconds duration is achieved. This pulse does both photolysis and probing of the Raman spectrum. Fig. 3.8 shows spectra of bacteriorhodopsin obtained from this apparatus. In each succeeding spectrum (top to bottom) the slit width is widened to increase the pulse duration. The spectral changes observed are, of course, complex. One must consider both the increase in illumination and the observation time, recognizing that a sum of the species present during this time is seen. Still, for production of microsecond pulses, the technique is a relatively simple one. The authors believe that 20 ns pulses can be obtained. However, one must be concerned about power-per-pulse, relaxation between pulses, and overall signal-to-noise considerations.

3.5 RAMAN OPTICAL ACTIVITY

Optically active molecules have a different scattering for left and right circularly polarized light. The theory of this effect, sometimes known as Raman circular intensity differential (CID) spectroscopy, was first outlined by Barron (71, 72).

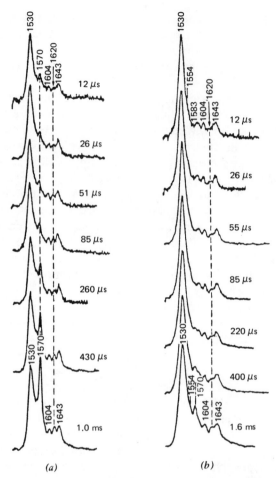

Figure 3.8 Kinetic resonance Raman spectra of bacteriorhodopsin taken with two excitation wavelengths: *(a)* 476.5 nm; *(b)* 514.5 nm; at room temperature (20°C). The times are the measured laser pulse durations, determined in general by the slit width, speed, and diameter of the chopper. For this series only the slit was varied. Peak incident laser power was 2 W with an average power (determined by the ratio of slit width to chopper circumference) ranging from 0.5 to 60 mW. Vibrational frequencies are given in cm^{-1}.

There have been several recent contributions to the theoretical predictions of this effect (73, 74). There is great interest in Raman CID as a possible method to assign absolute configurations to optically active centers to which it appears to be sensitive.

Experimental observations are quite difficult. Results of complete CID spectra for (−)-α-pinene and (+)-α-phenylethylamine were reported by Hug et al. (75). Hug and Surbeck (76) gave results for several substituted ethyl benzenes. A review by Nafie and Diem (77) provides a good introduction to the literature up to 1979.

The results presented thus far show that the Raman CID signal is usually about 10^{-3} to 10^{-4} times the total Raman signal. Given the weakness of the Raman effect itself, this is a very difficult measurement to make. All the sophisticated instrumentation techniques for excitation, polarization modulation, and detection have been used on this problem. It appears that the most likely possibility for signal improvement will come from nonlinear Raman techniques.

REFERENCES

1. L. J. Bellamy, *The Infrared Spectra of Complex Molecules,* 2nd ed., Methuen, London, 1958.

2. L. J. Bellamy, *Advances in Infrared Group Frequencies,* Methuen, London, 1968.

3. N. B. Colthup, L. H. Daly, and S. E. Wiberly, *Introduction to Raman and Infrared Spectroscopy,* Academic, New York, 1964.

4. C. N. R. Rao, *Chemical Applications of Infrared Spectroscopy,* Academic, New York, 1963.

5. R. N. Jones and C. Sandorfy, in *Chemical Applications of Spectroscopy,* Vol IX, W. West, Ed., Interscience, New York, 1956.

6. B. Schrader, *Angew. Chem.* (Int. Ed. Engl.) **12,** 884 (1973).

7. F. R. Dollish, W. G. Fateley, and F. F. Bentley, *Characteristic Raman Frequencies of Organic Compounds,* Wiley-Interscience, New York, 1974.

8. S. K. Freeman, *Applications of Laser Raman Spectroscopy,* Wiley-Interscience, New York, 1974.

9. G. Carlson, "Research and Development on the Analysis and Characterization of Experimental Materials," Report AFMA-TR-/-73-33, Wright-Patterson Air Force Base, Ohio, 1973.

10. R. A. Nyquist and R. O. Kagel, in *Infrared and Raman Spectroscopy,* Vol. 1, Part B, E. G. Brame and J. G. Grasselli, Eds., Dekker, New York, 1977, p 454.

11. W. Meier and B. Schrader, *Proc. Int. Conf. Raman Spectros.* **5,** 48 (1976).

12. H. A. Sloane, in *Polymer Characterization*, C. Craver, Ed., Plenum, New York, 1971.
13. P. Budinger, J. R. Mooney, J. G. Grasselli, P. S. Fay, and A. T. Guttman, *Anal. Chem.* **53**, 884 (1981).
14. W. Washburn, *Amer. Lab.* **10**:11, 47 (1978).
15. F. Behroozi, R. G. Priest, and J. M. Schnur, *J. Raman Spectrosc.* **4**, 379 (1976).
16. T. S. Wang and J. J. Mannion, *Appl. Spectrosc.* **27**, 27 (1973).
17. J. P. Lere-Porte, J. Petrissams, and S. Gromb, *J. Mol. Struct.* **34**, 55 (1976).
18. R. A. Nyquist, *Appl. Spectrosc.* **26**, 81 (1972).
19. S. K. Freeman and D. Mayo, *Appl. Spectrosc.* **27**, 286 (1973).
20. S. K. Freeman and D. Mayo, *Appl. Spectrosc.* **26**, 543, (1972).
21. C. Warren and D. Hooper, *Can J. Chem.* **51**, 3901 (1973).
22. B. Schrader and W. Meier, *Fresenius' Z. Anal. Chem.* **260**, 248 (1972).
23. J. N. Willis, R. B. Cook, and R. Jankow, *Anal. Chem.* **44**, 1228 (1972).
24. R. Oertel and D. Myhre, *Anal. Chem.* **43**, 974 (1971).
25. J. H. Van der Maas and T. Visser, *J. Raman Spectrosc.* **2**, 563 (1974).
26. T. Visser and J. H. Van der Maas, *J. Raman Spectrosc.* **6**, 114 (1977).
27. T. Visser and J. H. Van der Maas, *J. Raman Spectrosc.* **7**, 125 (1978).
28. T. Visser and J. H. Van der Maas, *J. Raman Spectrosc.* **7**, 278 (1978).
29. B. Schrader and E. Steigner, *Liebigs Ann. Chem.* **735**, 6 (1970).
30. E. Steigner and B. Schrader, *Liebigs Ann. Chem.* **735**, 15 (1970).
31. B. Schrader and E. Steigner, *Fresenius' Z. Anal. Chem.* **254**, 177 (1971); B. Schrader and E. Steigner, in *Modern Methods of Steroid Analysis*, E. Heftmann, Ed., Academic, New York, 1973.
32. M. M. Sushchinskii, *Raman Spectra of Molecules and Crystals*, Keter, New York, 1972, chap. 9.
33. F. A. Cotton, *Chemical Applications of Group Theory*, Wiley, New York, 1961.
34. H. G. Buge, P. Reich, and E. Steger, *J. Mol. Struct.* **35**, 175 (1976).
35. T. R. Manley and C. G. Martin, *Spectrochim. Acta, Part A* **32**, 357 (1976).
36. J. Maillois, L. Bardet, and L. Mavry, *J. Mol. Struct.* **30**, 57 (1976).
37. J. D. Lewis and J. L. Laane, *Spectrochim. Acta, Part A* **31**, 755 (1975).
38. W. Grossman, *Anal. Chem.* **46**, 345R (1974).
39. W. Grossman, *Anal. Chem.* **48**, 261R (1976).
40. D. Gardiner, *Anal. Chem.* **50**, 131R (1978).
41. D. Gardiner, *Anal. Chem.* **52**, 96R (1980).
42. L. A. Carreira, *J. Chem. Phys.* **62**, 3851 (1975).
43. L. A. Carreira and T. G. Townes, *J. Chem. Phys.* **63** 5283 (1975).
44. J. R. Durig, W. E. Bucy, and A. R. H. Cole, *Can. J. Phys.* **53**, 1832 (1975).

45. L. A. Carreira, *J. Phys. Chem.* **80,** 1149 (1976).
46. J. R. Durig and Y. S. Li, *J. Chem. Phys.* **63,** 4110 (1975).
47. J. R. Durig and A. W. Cox, Jr., *J. Chem. Phys.* **63,** 2303 (1975).
48. J. R. Durig, W. E. Bucy, C. J. Wurrey, and L. A. Carreira, *J. Phys. Chem.* **79,** 988 (1975).
49. J. R. Durig, A. C. Shing, L. A. Carreira, and Y. S. Li, *J. Chem. Phys.* **57,** 4398 (1972).
50. W. Kiefer, H. J. Bernstein, H. Wieser, and M. Danyluk, *J. Mol. Spectrosc.* **43,** 393 (1972).
51. W. Kiefer, H. J. Bernstein, M. Danyluk, and H. Wieser, *Chem. Phys. Lett.* **12,** 605 (1972).
52. C. Perchard and J. P. Perchard, *J. Raman Spectrosc.* **3,** 277 (1975).
53. N. L. Lavrik and Y. Naberulchin, *Opt. Spectrosc.* **37,** 44 (1974).
54. I. Pernall, U. Maier, R. Janoschek, and G. Zundel, *J. Chem. Soc., Faraday Trans. 2* **71,** 201 (1975).
55. O. H. Ellestad and P. Klaboe, *J. Mol. Struct.* **26,** 25 (1975).
56. H. T. Horntvedt and P. Klaboe, *Acta Chem. Scand., Ser. A* **29,** 427 (1975).
57. G. S. Kastha, S. D. Roy, and S. K. Nandy, *Indian J. Phys.* **46,** 293 (1972).
58. H. Matsuura, M. Hiraishi, and T. Myazawa, *Spectrochim. Acta, Part A* **28,** 229 (1972).
59. D. A. Compton, W. O. George, and W. Maddams, *J. Chem. Soc., Perkin Trans. 2,* 1666 (1976).
60. R. Wilbrandt, N. H. Jensen, P. Pagsberg, A. H. Sillesen, K. B. Hansen, and R. E. Hester, *Chem. Phys. Lett.* **60,** 315 (1979).
61. N. H. Jensen, R. Wilbrandt, P. Pagsberg, R. E. Hester, and E. Ernstbrunner, *J. Chem. Phys.* **71,** 3326 (1979).
62. K. B. Hansen, R. Wilbrandt, and P. Pagsberg, *Rev. Sci. Instr.* **50,** 1532 (1979).
63. E. Mayer, R. B. Girling, and R. E. Hester, *J. Chem. Soc., Chem. Comm.* 192 (1973).
64. L. R. Dosser, J. B. Pallix, G. H. Atkinson, H. C. Wang, G. Levin, and M. Szwarc, *Chem. Phys. Lett.* **62,** 555 (1979).
65. G. H. Atkinson and L. R. Dooser, *J. Chem. Phys.,* **72,** 2195 (1980).
66. H. Fabian, A. Lau, W. Werncke, and K. Lenz, *Chem. Phys. Lett.* **48,** 607 (1977).
67. W. F. Howard, Jr. and L. Andrews, *Inorg. Chem.* **14,** 767 (1975).
68. W. F. Howard, Jr. and L. Andrews, *J. Amer. Chem. Soc.* **97,** 2956 (1975).
69. E. S. Prochaska and L. Andrews, *Inorg. Chem.* **16,** 339 (1977).
70. A. Campion, M. A. El-Sayed, and J. Terner, *Biophys. J.* **20,** 369 (1977).
71. L. D. Barron and A. D. Buckingham, *Mol. Phys.* **20,** 1111 (1971).
72. L. D. Barron, *J. Chem. Soc.* **A,** 2900 (1971).

73. P. L. Prasad and D. F. Burrow, *J. Amer. Chem. Soc.* **101,** 800 (1979).

74. P. L. Prasad and L. A. Nafie, *J. Chem. Phys.* **70,** 5582 (1979).

75. W. Hug, S. Kint, G. F. Bailey, and J. R. Scherer, *J. Amer. Chem. Soc.* **97,** 5589 (1975).

76. W. Hug and H. Surbeck, *Chem. Phys. Lett.* **60,** 186 (1979).

77. L. A. Nafie and M. Diem, *Acc. Chem. Res.* **12,** 296 (1979).

Chapter Four

Polymers

A large number of publications, including some excellent review articles (1–5), discuss applications of using Raman spectroscopy to studies of polymers. One of the real advantages of using Raman spectroscopy in polymer characterization is the ease of obtaining a good spectrum with little or no sample handling. Injection-molded pieces, pipe and tubing, blown film, cast sheets, or monofilaments can be examined directly. If the thermal history is important for understanding the properties of a plastic, it is clearly undesirable to melt or dissolve the sample, as may be necessary for IR spectroscopy. Also, different properties are associated with the amorphous as compared to the more ordered regions within a polymer, and the Raman spectrum, obtained on the sample in its state as a finished product from which the physical property information is obtained, is of great value. Filled polymers, like composites, contain fillers such as glass or clay that strongly interfere with the IR spectra of the polymers because of their own intense absorption. On the other hand, glass and clay are poor Raman scatterers, so the Raman spectrum can be obtained without removal of the filler or any preparative techniques.

4.1 POLYMER CHARACTERIZATION

Raman spectroscopy is extremely valuable in many different areas of polymer characterization. In addition to the identification of the specific type of polymer, it can be used to determine functional groups, end-groups, structure, conformation, and orientation of chains, and to follow changes in structural parameters as the polymers are exposed to environmental or mechanical stresses (6). Table 4.1 rates the usefulness of Raman spectroscopy for various kinds of polymer analysis.

Raman spectroscopy has also been applied to studies of the mechanisms of

59

Table 4.1   Uses of Raman Spectroscopy

	Excellent	Very Good	Good	Poor
		Usefulness of Raman		
Polymer				
Homonuclear backbone	X			
Polar substituents			X	
End-groups				X
Multicomponent Systems				
Additives ($<$ 1%)				X
Fillers (5% $>$)				
Glass	X			
Carbon black				X
Inorganic (TiO_2)		X		
Pigments			Variable	
Properties				
Variable size and shape	X			
Limited solubility	X			
Colors with aging				X
Sensitive to thermal history	X			

polymer reactions. Koenig (6) followed the polymerization of butadiene, which can yield any of the different structures shown in Fig. 4.1. The polymerization conditions determine the amounts of 1,2 and 1,4 structures formed, which are important because the relative concentrations of each present in the end product determines its properties. Unsaturation cannot be measured by the C=C stretching band in the IR because the band is very weak (in fact, it is not observed for the 1,4 trans structure). However, the C=C stretch is the strongest band in the Raman spectrum, and the type of unsaturation can easily be determined by its position.

The impact of data systems on the applications of Raman spectroscopy has already been described (Section 2.1). Fig. 4.2 shows schematically the totally integrated system in the Sohio Molecular Spectroscopy Laboratory. Three ded-

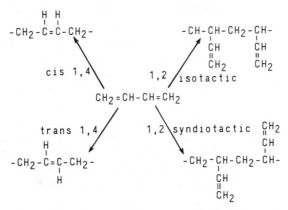

Figure 4.1 Structures resulting from butadiene polymerization. (From ref. 6.)

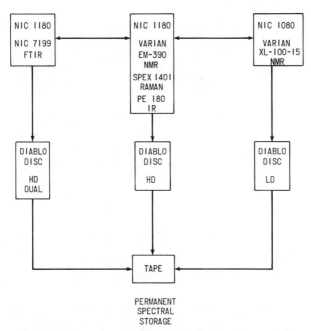

Figure 4.2 Integrated data system in Sohio molecular spectroscopy laboratory. (From ref. 7.)

61

icated minicomputers are utilized for instrument control, data acquisition, and data reduction of five instruments. RS-232 interfaces permit foreground, background operation and data output on any instrument recorder. Grasselli et al. describe an example illustrating the versatility of the system in following the extent of cross-linking in an experimental polystyrene polymer (7). The sample was insoluble in typical NMR solvents, and, because of the high sensitivity in the Raman for C=C stretching, Raman was selected over IR analysis as the analytical method. The powdered samples were packed into a capillary tube and the spectra were recorded over the region from 1500 to 1700 cm^{-1}. The Raman spectrum shown in Fig. 4.3 was plotted on the Nicolet FT-IR spectrometer. Its presentation is in IR format with peaks going down, rather than in Raman format with peaks going up. To follow the extent of cross-linking, the area of the C=C mode at 1645 cm^{-1} due to the cross-linking agent was ratioed to the area of a polystyrene ring mode at 1590 cm^{-1}, which was used as an internal standard band. The integration was performed using an NMR program on the Nicolet 1080 computer. The ratio of the two band areas for the starting material was compared to that for the product. The extent of cross-linking was followed very

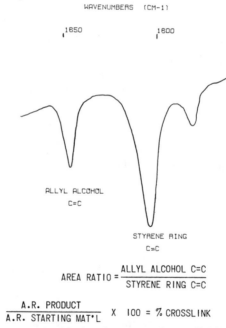

Figure 4.3 Raman spectrum showing the extent of cross-linking in an experimental polystyrene polymer. (From ref. 7.)

rapidly by watching the disappearance of the C=C band due to the cross-linking agent, allyl alcohol, as compared to the C=C polystyrene ring band.

In a similar example, the extent of cure in a polyester resin was measured by Raman spectroscopy. Again, Raman spectroscopy was utilized because of its high sensitivity and specificity to C=C bonds. The samples were polystyrene-based polyester resins and a method was established using the band at 1625 cm^{-1} due to the symmetric stretching vibration of the vinyl C=C in the monomer, styrene, and the 1590 cm^{-1} band due to the aromatic ring structure (7).

Other polymer compositional analyses have been done with Raman spectroscopy. Meeks and Koenig have made quantitative measurements on copolymers of vinyl chloride and vinylidene chloride (8, 9). Boerio and Yuann reported compositional analysis of styrene glycidal methacrylate and methyl methacrylate glycidalmethacrylate copolymers (10). Sloan and Bramston-Cook published quantitative results in a three component polymer system (11) containing styrene, butadiene, and methylmethacrylate. Mukherjee et al. (12) developed a method for the determination of terminal thiol groups in sulfur polymers using ethyloxyacetate as an internal standard.

4.2 POLARIZATION MEASUREMENTS OF POLYMERS

Raman polarization measurements on polymers can provide extremely valuable structural information, but they are sometimes difficult to obtain. Optical clarity and fluorescence can cause trouble, especially with impure or chemically complex samples. This is a particular problem for polarization studies, in which scrambling is caused by multiple scattering in a heterogeneous system (13). Solutions to these problems include a combination of careful sample handling and sophisticated measuring techniques.

The basic equipment necessary for polarization measurements on polymers has been described by Shepherd (14), Hendra (4), and Gilson and Hendra (15). More sophisticated equipment is necessary if any of the problems outlined above are serious, and this has been described by Shepherd (13).

Molecular conformation studies are extremely dependent upon accurate polarization measurements. Disordered polymers give vibrational modes that are all active in the IR and Raman, and all of the Raman lines are polarized. Polyvinyl fluoride is an example (9). Ordered structures show substantial differences in IR and Raman frequencies, therefore making it necessary to include Raman polarization and IR dichroic measurements to determine the conformation of the chain.

Jasse and co-workers (16) recently described the influence of the chain conformation on the ν_1 and ν_{13} normal modes of the benzene ring in atactic and isotactic polystyrene. A series of model compounds whose conformations were confirmed by NMR analysis were used to establish the band assignments. It was found that the ν_1 mode is influenced by the local conformation of the alkyl chain as well as by the length of the conformation structure along the chain. The ν_{13} mode is only influenced by the local conformation and is insensitive to the length of the alkyl chain.

Changes in conformation when polymers are put into aqueous solution can

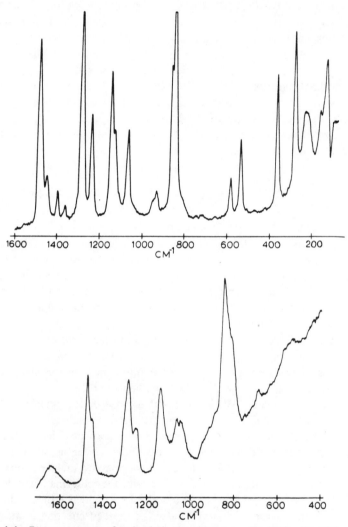

Figure 4.4 Raman spectra of polyethylene oxide: *(a)* solid PEO; *(b)* 10% aqueous solution of PEO. (From ref. 17.)

also be followed, since water is not a strong Raman scatterer. Koenig and co-workers (17) studied the structure of polyethylene oxide (PEO) in aqueous solution. Fig. 4.4 shows the spectra of crystalline PEO and a 10% water solution PEO. The crystalline spectrum is complex, showing several sharp bands with characteristic helical splitting. The spectrum of the melt has broader bands at shifted frequencies, and no band splitting is detected. This indicates the helical structure of the solid polymer is lost upon melting. On the other hand, the spectrum of PEO in water looks much more like the crystalline polymer, indicating that considerable helical structure is left upon dissolution.

Koenig (9) has summarized selection rules for Raman and IR activity for monosubstituted vinyl polymers (Fig. 4.5). These polymers can exist in a number of configurations. They may be syndiotactic or isotactic in a helical or planar conformation. Alternatively, they can be atactic with a disordered conformation. Each of these structures has its own selection rules. With the aid of such a spectroscopic scheme, ideally, it should be possible to unequivocally determine the structure of any unknown monosubstituted vinyl polymer.

Bailey et al. have used polarization data to determine mode assignments for isotactic polypropylene (IPP) (18). The uniaxially oriented polymer was studied in a variety of geometries, and some of these data are shown in Fig. 4.6. Significant degrees of polarization were observed in spite of the incomplete orientation and optical scrambling. IPP, which has a 3_1 helical form (a helix

STRUCTURE	SYMMETRY	R: p / IR: π	p / σ	d / π	d / σ	p / O	d / O	O / π	O / σ	EXAMPLE
CENTER OF SYMMETRY	D_{2h} C_{2h}					√	√	√	√	PE, PES
ATACTIC		√	√	√	√					PVF
HELIX $>3_1$ (syndiotactic)	D_n				√	√	√	√		PEO
HELIX 3_1 (syndiotactic)	D_3				√	√	√			
HELIX 2_1 (syndiotactic)	D_2			√	√	√				
PLANAR (syndiotactic)	C_{2v}	√	√	√		√				PVC
HELIX $>3_1$ (isotactic)	C_n	√		√		√				POLYBUTENE
HELIX 3_1 (isotactic)	C_3	√		√						PP
PLANAR (isotactic)	C_S	√	√							

Figure 4.5 Selection rules for monosubstituted vinyl polymers. (From ref. 9.)

65

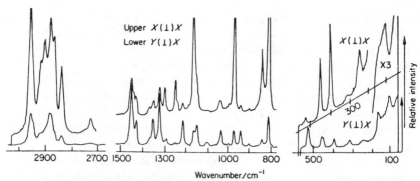

Figure 4.6 Raman spectra of oriented isotactic polypropylene. The orientation axis is perpendicular to both the observation axis Y and the laser beam direction Z. In each spectrum the scattered polarizaton is X; the incident polarization is X (upper) or Y (lower). (From ref. 18.)

with 3 monomers per turn), can be treated as belonging to a group isomorphous with the point group C_3; 25 normal vibrations can be classified as A modes and 52 as B modes. Both these species can be distinguished in the spectra, and the assignments agree well with previous IR (19) and Raman results (20).

The Raman spectrum of the two-fold helix of syndiotactic polypropylene has also been obtained and interpreted (21). Chalmers found good agreement between observed and predicted shifts, which verified the force field and proposed structure of Schachtschneider and Snyder.

Another important area of research in polymer studies involves how the physical and mechanical properties of a polymer are influenced by molecular orientation induced by drawing. Raman polarization studies can give detailed information about the distribution of orientations of structural units for both crystalline and noncrystalline regions. This has been evidenced in work on oriented poly (methylmethacrylate) (PMMA) (22, 23) and poly (ethylene terephthalate) (PET) (22, 24). Hendra and Willis (25, 26) have studied oriented polypropylene and polyethylene; Derouault et al. have studied inhomogeneous PET (27), and Gall et al. (28) have worked with polyethylene (PE). None of these latter efforts included quantitative estimates of orientation.

Quantitative calculations of molecular orientation have been made for PE, and these have been summarized by Shepherd (13) in Table 4.2. The studies used two PE "crystalline" bands at 1170 and 1296 cm^{-1} and one "amorphous" band at 1081 cm^{-1}. Values of the averages $\cos^2 \theta$ and $\cos^4 \theta$ were calculated, where θ is the angle between a unique axis in the unit and the draw direction. IR results for $\cos^2 \theta$ are also included in the table, and the agreement with the Raman

Table 4.2   Orientation Functions for Oriented Polyethylene Films[a]

Band	Assignment	From Raman Data		From Infrared Data	Theoretical	
ν/cm^{-1}		$\cos^2 \theta$	$\cos^4 \theta$	$\cos^2 \theta$	$\cos^2 \theta$	$\cos^4 \theta$
1081	Amorphous (gauche isomer)	0.415	0.334	0.45	0.45[b]	0.31[b]
1170	$a_g + b_{1g}$	0.831	0.763	0.85	0.83[c]	0.74[c]

[a]J. Maxfield, R. S. Stein, and M. C. Chen, *J. Polym. Sci., Polym. Phys. Ed.* **16,** 37 (1978).
[b]Based on ref. 29.
[c]Based on ref. 30.

measurements is good. Theoretical calculations were also made using a rubber elasticity model for the amorphous band (29) and a spherulitic model (30) for the crystalline averages. Again, the agreement is surprisingly good.

Quantitative data were also obtained on chain orientation in hydrostatically extruded polypropylene by Satija and Wang (31). The alignment of the polymer chain along the direction of extrusion was found to increase with the increase of the extrusion ratio. The subject of vibrational spectroscopic studies of polymer chain order has been reviewed (32).

4.3   MORPHOLOGICAL EFFECTS

Morphological effects in polymers can readily be studied by Raman spectroscopy. Many polymers solidify from the melt in spherulites composed of lamellar structural units. Measurement of the detailed structure of lamellae is of great interest to polymer chemists because of its relationship to mechanical strength and stability of the material (33, 34). Although x-ray diffraction, electron microscopy, and differential scanning calorimetry are commonly used to obtain information on crystallinity and lamellar thickness, Raman spectroscopy has also been proposed as a convenient and rapid technique for measuring the dimensions of chain-folded lamella.

A low frequency Raman band in *n*-paraffins has a frequency that is inversely proportional to chain length. The band was assigned to the fundamental longitudinal acoustic mode (LAM) of the chain and was observed in PE, poly (ethylene oxides), and polypropylene (35–38). The LAM was used by some workers to measure "chain length" and lamellar thickness, but an elegant theoretical treatment of the LAM, published by Krimm and co-workers (35–38), shows that the LAM frequencies are affected by perturbing forces in the ends of the chains as well as by relative moduli, densities, and fractions of amorphous and crystalline components. Nevertheless, Raman spectroscopy is extremely helpful in determining the mechanism of lamellar formulation, the structure of fold sequences, and the nature of interlamellar adhesive forces (39, 40). Deformation and cooling effects on the lamellar structure of high density PE have already been studied (41).

4.4 OTHER POLYMER FEATURES

In addition to conformation and configuration, Raman spectra are sensitive to other polymer features such as the degree of crystallinity, chain folding, molecular weight, and end-groups (42). Some of the spectral changes associated with these factors can be seen in Fig. 4.7, which shows the Raman spectra of various forms of PE. Consideration of molecular weight, degree of crystallinity, etc., is important because of the relationship of these factors to physical properties such as hardness, brittleness, permeability, and impact resistance.

In an interesting study of styrene-sodium methacrylate copolymers, Raman spectroscopy was used to determine the concentration of ion pairs in multiplets and clusters (43).

4.5 DETECTION OF IMPURITIES IN POLYMER FILMS

Raman spectroscopy is also a very effective technique for examining impurities in polymer films (34). If caution is taken to prevent thermal decomposition of the sample, the laser beam can be condensed with a lens to examine specks of material with dimensions on the order of 10 μm. By contrast with IR spectroscopy, there is no problem with a beam that has not impinged on the speck entering the monochromator, since the scattered light is collected at either 90° or 180° to the axis of the exciting beam. Also, there is no distortion of the Raman spectrum if the sample is irregular in shape or thickness.

Fig. 4.8 shows the spectrum of an ethylene vinyl acetate (EVA) film containing a small speck of foreign material suspected of being another polymer. Also included in the figure is a reference of "pure" EVA and two suspected impurities—low and high density polyethylene (LDPE and HDPE, respectively). Although the spectra are generally very similar, small differences can be noted. For example, the HDPE has a band at 1640 cm^{-1}, assigned to terminal vinyl —C=CH$_2$ groups, that does not appear in the LDPE. The spectrum of the impurity also shows unsaturation, but it occurs at 1660 cm^{-1} and is assigned to trans —CH=CH. This type of unsaturation occurs when EVA copolymer is

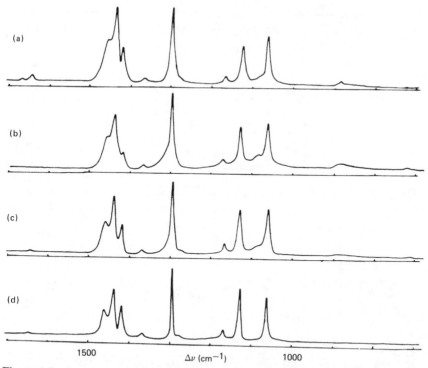

Figure 4.7 Raman spectra of some polyethylenes: *(a)* Raman spectrum of a wax lump with molecular weight approximately 800, which melts at ca. 75°C; *(b)* spectrum of unoriented film; *(c)* and *(d)* spectra of high density, high melting point forms with molecular weight 10^4 to 10^5. The intensity of the feature at 890 cm^{-1} is a rough inverse indicator of the molecular weight. Crystallinity, a desirable characteristic from the standpoint of oxidative degradation, is indicated by the sharpness of the bands at 1300 cm^{-1} and 1400 to 1500 cm^{-1}. Thus *(b)*, with a crystallinity of 50%, exhibits the widest Raman lines; *(d)*, with almost total crystallinity (the extended chain form), shows the narrowest lines. When molecular chains terminate in a vinyl grouping, the features around 1650 cm^{-1} appear. (From ref. 42.)

heated and acetic acid evolves. Thus the impurity appears to be due to thermal decomposition of the polymer and not to either type of PE.

4.6 MISCELLANEOUS POLYMER APPLICATIONS

There have been many applications of Raman spectroscopy to studies of specific polymers or types of polymers. In rubber chemistry, Coleman et al. measured

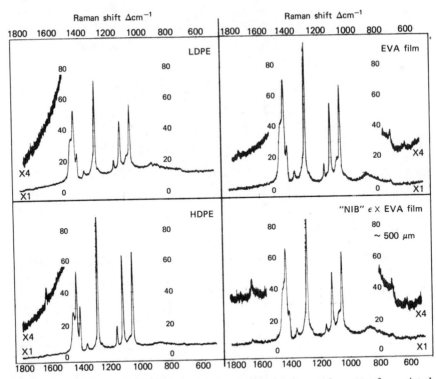

Figure 4.8 Raman spectrum of impurity from a polymer film, and spectra of associated materials. (From ref. 34.)

Raman spectra of vulcanizates prepared from *cis*-1,4-polybutadiene, 2-mercaptobenzothiazole, sulfur, zinc oxide, and lauric acid (44). A technical review gives evidence from Raman spectroscopy supporting a mixed free radical and ionic mechanism for accelerated sulfur vulcanization (45). In a separate publication, Coleman et al. also assigned Raman bands to the crystalline and amorphous components of chloroprene polymers (46). Crystallinity was studied in ethylene-propylene rubbers by Schreier and Peitscher (47).

Several acrylic polymers have been examined by Raman spectroscopy. In addition to the orientation studies on PMMA already mentioned (22, 23), the low frequency spectrum has been obtained and discussed (48). The neutralization of syndiotactic polyacrylic acid in aqueous solution was examined by Bardet et al., and a planar zig-zag configuration was observed for the nonionized acid and its sodium salt (49).

Maddams and co-workers (50, 51) have used Raman spectroscopy in structural studies on polyvinylchloride. They obtained more complete assignments of the carbon-chlorine stretching modes through detailed analysis of band positions and band shapes, using computer resolution of overlapping bands to refine the information content. Of equal interest, analytically, was their characterization of polyene sequences between 9 and 17 conjugated double bonds in degraded polyvinylchloride samples. Koenig extended this work using the 363.9 nm laser line to detect apparent sequences of 7 to 8 conjugated double bonds in polyvinylchloride (52). Certainly the characterization of such sequences in degraded polymers suggests new possibilities, both in detecting the initiation of degradation from the appearance of short conjugated sequences and in measuring the longer sequences which cannot be done by UV-visible spectroscopy. It is possible, too, that as the tacticity of the polymer changes, the degradation mechanism will be affected. Studies of this nature are underway (53).

The effect of cross-linking on the Raman spectra of epoxy resins has been studied by Lu and Koenig (54) and on unsaturated polyester resins containing styrene by Koenig and Shih (55). In the hopes of obtaining some insight into the fracture mechanism of polymers, the effect of mechanical stress on the Raman spectra of polypropylene, polycarbonate, polystyrene, and nylon 6,6 has been examined (56).

The structural changes that occur in wool after annealing have also been investigated (57) by Shishoo and Lundell. Results show that the CH_2 groups in wool have parallel polarization in the Raman spectrum, and the number increases significantly after annealing. The authors have formulated the hypothesis that a large number of electrovalent intrahelix cross-linkages exist between the suitable side groups on the main polypeptide chain in the helix of annealed wool.

4.7 FUTURE OF RAMAN POLYMER STUDIES

The future of Raman polymer studies appears to hold great potential for many different applications. Improved force fields should contribute to a better understanding of regular and irregular conformations of polymers (58). Polymer crystal studies should add to the existing knowledge of intramolecular forces in crystals and their effect on stable polymer structures. Intensity studies should also be valuable. There has been great progress in use of integrated optics techniques for studying thin films (59). Perhaps most important, however, will be the increased use of vibrational spectroscopy to understand the relationship between structure and function, an understanding that could have an important impact on designing polymers for specific physical properties.

REFERENCES

1. J. L. Koenig, *Appl. Spectrosc. Rev.* **4**, 233 (1971).
2. W. L. Peticolas, *Adv. Polym. Sci.* **9**, 285 (1972).
3. B. G. Frushour and J. L. Koenig, in *Advances in Infrared and Raman Spectroscopy*, Vol. 1, R. J. H. Clark and R. Hester, Eds., Heyden, London, 1975.
4. P. J. Hendra, in *Polymer Spectroscopy*, D. O. Hummel, Ed., Verlag Chemie, Weinheim/Bergstr., 1974.
5. H. A. Willis, *Proc. 5th Eur. Symp. Polym. Spectrosc.*, Hummel, Weinheim, 1979.
6. J. L. Koenig, *Chem. Technol.* **2**, 411 (1972).
7. J. G. Grasselli, M. A. S. Hazle, J. R. Mooney, and M. Mehicic, *Proc. 21st Colloq. Spectrosc., Int.*, Heyden, London, 1979.
8. M. Meeks and J. L. Koenig, *J. Polym. Sci., Polym. Phys. Ed.* **9**, 717 (1971).
9. J. L. Koenig, *Appl. Spectrosc. Rev.* **4**, 233 (1971).
10. F. J. Boerio and J. K. Yuann, *J. Polym. Sci., Polym. Phys. Ed.* **11**, 1848 (1973).
11. H. Sloane and R. Bramston-Cook, *Appl. Spectrosc.* **27**, 217 (1973).
12. S. K. Mukherjee, G. D. Guenther, and A. K. Bhattacharya, *Anal. Chem.* **50**, 1591 (1978).
13. I. W. Shepherd, in *Advances in Infrared and Raman Spectroscopy*, Vol. 3, R. J. H. Clark and R. Hester, Eds., Heyden, London, 1977.
14. I. W. Shepherd, *Rep. Prog. Phys.* **38**, 575 (1975).
15. T. R. Gilson and P. J. Hendra, *Laser Raman Spectroscopy*, Wiley-Interscience, New York, 1970.
16. B. Jasse, R. S. Chao, and J. L. Koenig, *J. Raman Spectrosc.*, **8**, 244 (1979).
17. J. L. Koenig and A. C. Angood, *J. Polym. Sci., Polym. Phys. Ed.* **8**, 1787 (1970): F. J. Boerio and J. L. Koenig, *J. Polym. Symp.* **43**, 205 (1973).
18. R. T. Bailey, A. J. Hyde, and J. J. Kim, *Spectrochim. Acta, Part A* **30**, 91 (1973).
19. H. Tadakoro, M. Kobayashi, M. Ukita, K. Yasufuku, and S. Murahashi, *J. Chem. Phys.* **42**, 1432 (1975).
20. P. D. Vasko and J. L. Koenig, *Macromolecules* **3**, 597 (1970).
21. J. M. Chalmers, *Polymer* **18**:7, 681 (1977).
22. D. I. Bower, *Structure and Properties of Oriented Polymers*, Applied Sciences, London, 1975.
23. J. Purvis and D. I. Bower, *Polymer* **15**, 645 (1974).
24. J. Purvis and D. I. Bower, *J. Polym. Sci., Polym. Phys. Ed.* **14**, 1461 (1976).
25. P. J. Hendra and H. A. Willis, *Chem. Ind.* (London), 2146 (1967).
26. P. J. Hendra and H. A. Willis, *Chem. Comm.*, 225 (1968).
27. J. L. Derouault, P. J. Hendra, M. Cudby, and H. A. Willis, *J. Chem. Soc., Chem. Comm.*, 1187 (1972).

28. M. J. Gall, P. J. Hendra, C. J. Peacock, M. E. A. Cudby, and H. A. Willis, *Spectrochim. Acta, Part A* **28,** 1485 (1972).

29. R. J. Roe and W. R. Krigbaum, *J. Appl. Phys.* **25,** 2215 (1964).

30. D. Y. Yoon, C. Chang, and R. S. Stein, *J. Polym. Sci., Polym. Phys. Ed.* **12,** 2091 (1974).

31. S. K. Satija and C. H. Wang, *J. Chem. Phys.* **69:**6, 2739 (1978).

32. K. Holland-Moritz, *J. Appl. Polym. Sci. Appl. Polym. Symp.* **34** (Polym. Anal.), 49 (1978).

33. P. J. Hendra, in *Structural Studies of Macromolecules by Spectroscopic Methods,* K. J. Ivin, Ed., Wiley, London, 1976.

34. H. A. Willis, in *Molecular Spectroscopy,* A. West, Ed., Heyden, New York, 1977.

35. S. L. Hsu and S. Krimm, *J. Appl. Phys.* **47,** 4265 (1976).

36. S. L. Hsu, G. W. Ford, and S. Krimm, *J. Polym. Sci., Polym. Phys. Ed.* **15,** 1769 (1977).

37. S. L. Hsu and S. Krimm, *J. Appl. Phys.* **48,** 4013 (1977).

38. S. L. Hsu and S. Krimm, *J. Polym. Sci., Polym. Phys. Ed.* **16,** 2105 (1978).

39. G. R. Strobl, *Colloid Polym. Sci.* **257,** 584 (1979).

40. A. Peterlin, *J. Mater. Sci.* **14,** 2994 (1979).

41. G. V. Frazer, P. J. Hendra, M. E. Eudby, and H. A. Willis, *J. Mater. Sci.* **9,** 1270 (1974).

42. M. J. Gall and P. J. Hendra, *The Spex Speaker* **16,** 1, (1971).

43. A. Neppel, I. S. Butler, and A. Eisenberg, *J. Polym. Sci., Polym. Phys. Ed.* **17,** 2145 (1979).

44. M. M. Coleman, J. R. Shelton, and J. L. Koenig, *Rubber Chem. Technol.* **45,** 173 (1972).

45. M. M. Coleman, J. R. Shelton, and J. L. Koenig, *Ind. Eng. Chem. Prod. Res. Dev.* **13,** 154 (1974).

46. M. M. Coleman, P. C. Painter, and J. L. Koenig, *J. Raman Spectrosc.* **4,** 417 (1976).

47. G. Schreier and G. Peitscher, *Fresenius' Z. Anal. Chem.* **258,** 199 (1972).

48. S. J. Spells and I. W. Shepherd, *J. Chem. Phys.* **66,** 1427 (1977).

49. L. Bardet, G. Cassanas-Fabre, and M. Alain, *J. Mol. Struct.* **1,** 153 (1975).

50. D. L. Gerrard and W. F. Maddams, *Macromolecules* **8,** 54 (1975).

51. W. F. Maddams, *J. Macromol. Sci. Phys.* **14,** 87 (1977).

52. J. L. Koenig, Case Western Reserve University, Cleveland, Ohio, personal communication.

53. W. F. Maddams, The British Petroleum Co., Ltd., Chertsey Rd., Sunbury-on-Thames, Middlesex TW 16 7LN, U.K., personal communication.

54. C. S. Lu and J. L. Koenig, *Amer. Chem. Soc., Div. Org. Coat. Plast. Chem.,* Pap. 32, 112 (1972).

55. J. L. Koenig and P. Shih, *J. Polym. Sci., Polym. Phys. Ed.* **10,** 721 (1972).
56. R. A. Evans and H. E. Hallam, *Polymer* **17,** 838 (1976).
57. R. Shishoo and M. Lundell, *J. Polym. Sci., Polym. Chem. Ed.* **14,** 2535 (1976).
58. F. J. Boerio and J. L. Koenig, *J. Macrolmol. Sci. Rev. Macromol. Chem.* **7**:2, 246 (1972).
59. J. F. Rabolt, R. Santo, and J. D. Swalen, *Appl. Spectrosc.* **33,** 549 (1979).

Chapter Five

Biological Systems

An essential problem of modern biophysics is obtaining detailed structural information in solution or in phases of intermediate fluidity, such as membranes. In such problems Raman spectroscopy is emerging as an important tool, particularly when used in combination with other techniques. Table 5.1 summarizes some of the positive aspects of the application of Raman spectroscopy to biological samples. Only limited use has been made to date of isotopic substitution, but this will become more common for elucidation of assignments. Small sample sizes are a very important advantage of Raman spectroscopy over other solution techniques.

Table 5.1 also gives the negative aspects of applying the Raman effect to these problems. Many of these are general to all applications of Raman spectroscopy. Of these, fluorescence has thus far been the major problem, but the indirect nature of the conclusions obtained from Raman work may prove to be the ultimate limitation on its applicability.

5.1 PEPTIDES AND PROTEINS

Vibrational spectroscopy has long been used for studying conformations of peptides and proteins. The so-called amide I and amide II bands were widely used in IR spectra but were not readily observable in aqueous solution. In Raman spectra, primary use has been made of the amide I and amide III modes. The amide II band is generally very weak in the Raman spectrum. Fig. 5.1 shows an approximate description of these modes. The frequencies of these modes are sensitive to the angles ψ and ϕ defined in Fig. 5.2. Since these angles have characteristic values for the well-known α helix and β sheet protein structures and a range of values for random coils, the modes are diagnostic for protein

Table 5.1 Application of Raman Effect to Biological Problems

Positive Aspects

Spectra may be obtained in all phases
Aqueous solutions may be used
Sensitive to conformational change
Isotopic substitution affects spectra
Small sample size (5 nl volume)

Negative Aspects

Raman effect is weak
Fluorescence is $\sim \times 10^6$ as strong
Only moderate structural resolution
Conclusions often rely on indirect methods
Photochemistry may interfere

conformation in solution. Table 5.2 shows a summary of frequencies and intensities expected for various structures, and Table 5.3 gives some specific examples of modes that have been observed for various peptides.

Poly-L-lysine spectra are a good example of the kind of information that can be obtained. Fig. 5.3 shows the spectra of α helical and β sheet forms of this peptide obtained by Peticolas (1). These are in reasonably dilute aqueous solution. The amide I mode, concealed by a strong water band in the α helical form, appears strongly in the β sheet. Changes are also seen in the amide III region and the C—C stretching modes in the 945 to 1000 cm^{-1} region. One question related to these conformations that was answered by Raman spectroscopy was whether in transforming from $\alpha \rightarrow \beta$ the polymer goes through an intermediate structure, for example, a random coil. This was not the case. A quantitative plot of the temperature dependence of the amide III intensity at 1240 cm^{-1} relative to another band in the spectrum allows the determination of the fraction of either conformer present.

In proteins, there is apt to be a distribution of conformers present in a single system. For such a case, the spectra are necessarily more complicated. Lord (2) has proposed that the intensity contour in the amide III region might be directly correlated with the distribution of ψ values for the enzyme, as shown in Fig. 5.4. More work is needed to establish the generality of the correlation.

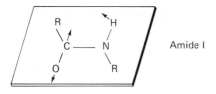

Amide I

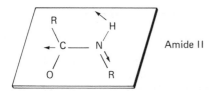

Amide II

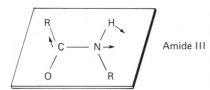

Amide III

Figure 5.1 There are many different modes of the peptide bond vibration. The figure illustrates the "in-plane" vibrational modes of the peptide bond. Among the three modes shown in the figure, the amide I and III bands are indices of peptide backbone conformation of a protein in the Raman spectra. The amide II band is either Raman inactive or very weak.

An approach related to Lord's has been taken by Lippert et al. (3). They suggest that four simultaneous equations can be used to establish quantitatively the fractions of α helix, β sheet, and random coil in a protein. To do this, they use the Raman intensities at 1240 cm^{-1} in H$_2$O solution and 1632 and 1660 cm^{-1} in D$_2$O, relative to the intensity of CH$_2$ deformation at 1448 cm^{-1}. Using poly-L-lysine as a standard to set the scale, they apply this method to nine proteins with good results ($\pm$ 10 to 15%). Craig and Gaber (4) have applied this method to a structurally well-documented enzyme, human carbonic anhydrase B, with good results as well.

Figure 5.2 Definition of the torsional angles ϕ and ψ about the C$^\alpha$—N bond and the C$'$—C$^\alpha$ bond. In the extended form shown, $\phi = +180°$, $\psi = 180°$.

77

Table 5.2   Conformation-Sensitive Raman Lines of Peptidyl Groups of Aqueous Proteins

$\sigma(cm^{-1})$	Assignment	Relative Intensity	Conformational Significance
1665–1672	Amide I	Strong	β sheet
1660–1670	Amide I	Strong, broad	Random chain
1645–1655	Amide I	Strong	α helix
1270–1300	Amide III	Weak	α helix
1243–1253	Amide III	Medium, broad	Random chain
1229–1235	Amide III	Strong	β sheet

To maintain conformation of proteins, disulfide bonds are often built into the structure. The conformation about these disulfide bonds is often a crucial piece of information to be obtained in elucidating protein structure. Raman spectroscopy has proved to be useful for studying this problem. The polarizable sulfur atoms tend to give rather intense Raman scattering. Fig. 5.5 shows three possible cases and the S—S stretching frequencies observed for each. Fig. 5.6, a spectrum of a neurotoxin obtained from sea snake (5), illustrates an application of this by Yu and co-workers, showing that the conformation is gauche-gauche-gauche. In a series of papers, Yu, Tu, and co-workers (6, 7) have shown that the snake neurotoxins are all primarily β sheet structures, with gauche-gauche-gauche S-S linkages. Approximate molecular orbital methods indicate that there is a

Table 5.3   Amide III Frequencies in Polypeptides from Raman Measurements

Substance	Amide III Frequency (cm^{-1})		
	α helical	β structure	Random Coil or Structure Ionized
Polyglycine	1261	1234	
Poly-L-alanine	1261	1239	
Poly-L-glutamic acid			1248
Glucagon	1266	1232	1248
Poly-L-lysine	1311	1240	1243
Poly Ala-Gly	1271	1238	
Poly Ser-Gly	1264	1236	

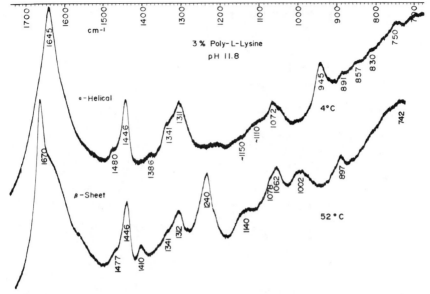

Figure 5.3 Raman spectra: upper, α helical poly-L-lysine (3% H_2O, pH 11.8, $T = 4°C$); lower, β sheet poly-L-lysine (3% H_2O, pH 11.8, $T = 52°C$). (From ref. 1.)

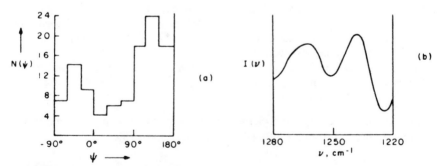

Figure 5.4 *(a)* Histogram of distribution of values of ψ in the range $-90°$ to $+180°$ in native bovine pancreatic ribonuclease; *(b)* intensity contour of the amide III region in the Raman spectrum of the enzyme. (From ref. 2.)

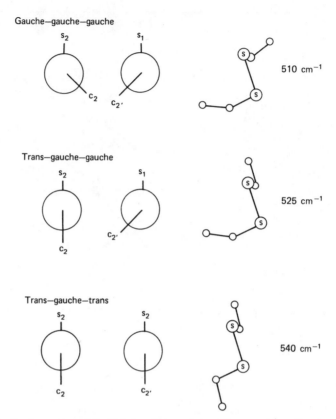

Figure 5.5 Conformation of disulfide bonds in proteins. The stretching vibration of S—S is influenced by the rotational conformation about the C—C and C—S bonds. The assignment numbers for C and S atoms are: C_2-$C_1$$S_1$-$S_2$-$C_1$-$C_2$. The disulfide bond with gauche shows a strong and sharp band at 510 cm^{-1}. The disulfide bonds with trans-gauche-gauche and trans-gauche-trans conformation give bands at 525 and 540 cm^{-1} in Raman spectra. (From ref. 5.)

basis for the correlation of S-S frequencies with dihedral angle in the orbital overlaps obtained from a Mulliken population analysis (8).

5.2 NUCLEOTIDES AND NUCLEIC ACIDS

While some work on peptides, amino acids, and proteins had been carried out using Raman spectroscopy prior to the laser era (that of Edsall and co-workers is particularly notable), the spectra of nucleic acids and polynucleotides is more recent. A wide variety of problems have been studied by Raman spectroscopy with impressive results. Table 5.4, from Thomas (9), indicates the range of problems that have been investigated with some success. These include constitution, secondary structure, and dynamics. There have also been more sophisticated applications to protein-DNA or -RNA interaction.

These studies illustrate well the importance of Raman spectroscopy as a technique for studying solution structures where the solid state structures have already been elucidated using crystallographic methods. There have been several extensive reviews of this work (10, 11).

Thomas and co-workers (12) have carried out considerable work on the model systems poly (A), poly (U), poly (G), and poly (C), which can exist in double helical forms such as poly (A) · poly (U). Fig. 5.7 shows a spectrum of poly (A) · poly (U) in D_2O. There is a striking temperature dependence of this

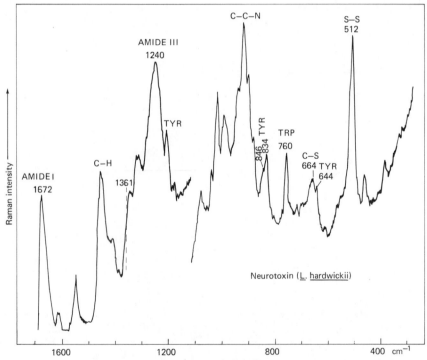

Figure 5.6   Raman spectrum of purified sea snake neurotoxin. (From ref. 5.)

81

Table 5.4   Nucleic Acid Problems Studied by Raman Spectroscopy

Primary Structures of Nucleic Acid Constituents

Tautomeric equilibria of the purine and pyrimidine bases
Ionization equilibria of the bases
Ionization equilibria of the phosphate groups
Sites of deuterium-hydrogen exchange in base and sugar residues

Secondary Structures and Interactions of Nucleic Acids

Mode and extent of hydrogen bonding between bases
Mode and extent of base-stacking interactions
Identification of ordered and disordered backbone conformations
Order-disorder transitions as a function of temperature, ionic strength, and so on
Quantitative estimates of RNA secondary structure
Binding of metal ions to base and phosphate sites
Nucleic acid-protein interactions
Hydration of nucleic acids

Kinetics

Rate of exchange of 8-CH in purines
Rate of hydrolysis of nucleic acids

From ref. 9.

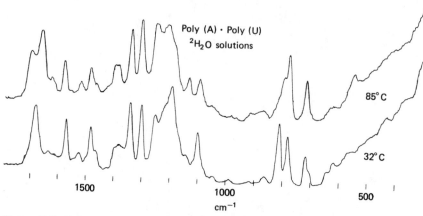

Figure 5.7   Raman spectra of D_2O solutions of poly (A) · poly (U) at 32 and 85°C.
Solute concentration is 35 mg/ml. (From ref. 12.)

82

spectrum as the double helix dissociates to form two single stranded helices. This illustrates the sensitivity of Raman spectroscopy to such changes.

Fig. 5.8 shows spectra of tRNA taken in both H_2O and D_2O. Assignments of the various bands to nucleotide residues are indicated in the figure. Using an empirical assignment approach, it is possible to synthesize such spectra with some confidence from polynucleotide spectra fragments. The use of D_2O allows one to examine vibrations obscured by H_2O scattering, particularly in the 1650 cm^{-1} region. It also permits study of the exchangeable protons in the RNA. Chan and Thomas (13) have used this method to establish the kinetics of such exchange in one case.

In simple viruses, there exists a so-called coat protein that encapsulates RNA or DNA. The coat protein actually may consist of many protein molecules. Combining the work on protein spectra and nucleic acid spectra, the interactions between the nucleic acid and the protein have been studied by Raman spectroscopy.

Fig. 5.9 shows the spectra of two DNA virus particles, known as Pf1 and fd virions (14). High quality spectra are obtainable even from such complex systems. Because these particles are very high in protein content, this Raman scattering dominates the spectra. From the spectra one can learn a great deal about these systems. Drawing on correlations previously presented, one notes that both contain almost exclusively α helical protein; on the other hand, the amino acid compositions are seen to be quite different for the two cases. It is also possible to conclude that OH groups on the tyrosyl amino acid residues are strongly hydrogen bonded to positive donor groups in both cases. This is gleaned from the relative intensity of bands at 854 and 325 cm^{-1}. From model system work, it is possible to conclude that the DNA backbones are of the β helix type rather than the α helix. These brief summaries of conclusions presented in great detail in the original work give an idea of the level of information that can be extracted from the spectra.

5.3 RESONANCE RAMAN SPECTRA OF BIOLOGICAL SYSTEMS

Conventionally, Raman spectra require relatively high sample concentrations on the order of 0.1 M or greater. This may limit their application to real biological problems. However, when the Raman exciting line falls within the envelope of an electronic absorption band, it becomes possible to see spectra at concentrations of approximately 10^{-5} M. Qualitatively we may say that these spectra arise only from the part of the molecule associated with the electronic transition. Thus in the case of a protein-like hemoglobin containing a chromophore (heme) that is

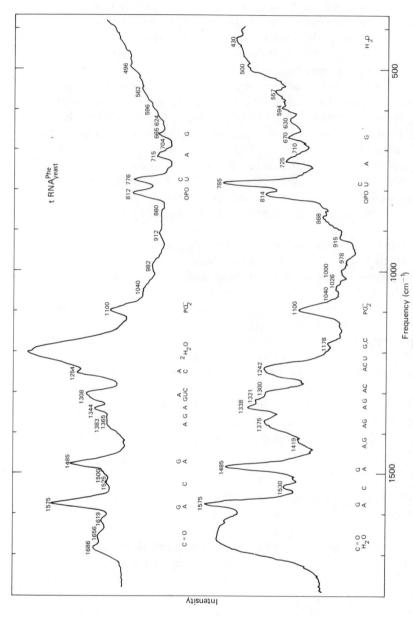

Figure 5.8 Raman spectra of tRNA $^{Phe}_{yeast}$ obtained with 488.0 nm excitation: (a) H_2O solution, 40 µg/µℓ, pH 7, 32°C; (b) D_2O solution, 40 µg/µℓ, pH 7, 32°C. Frequencies of the prominent lines are listed in cm^{-1} units and assignments to base and backbone residues are indicated by letter symbols along the abscissa. (From ref. 13.)

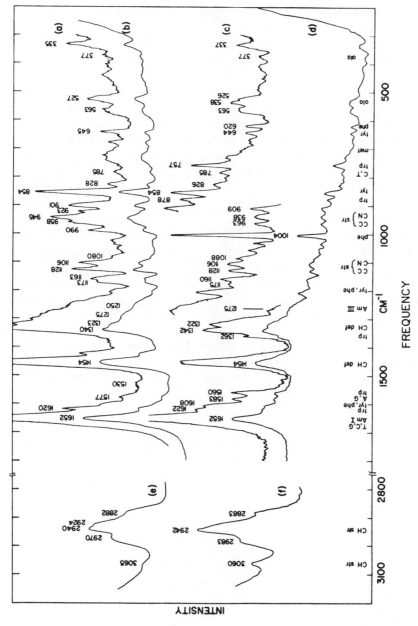

Figure 5.9 Raman spectra of FB viruses in 0.05 M NaCl at pH 9 and 32°C. (From ref. 14.)

orbitally isolated from the bulk of the globin protein, we see a selectively enhanced spectrum.

Pioneering work on this problem was carried out by Spiro and Strekas (15). They published the first spectra of resonance enhanced hemoglobin and cytochrome C. A particularly interesting feature was first observed in these spectra, although it had been predicted by Placzek, and other workers had attempted to observe it in model compounds without success. When the symmetry is either D_{4h} or slightly reduced from D_{4h}, certain modes have extremely large depolarization ratios. While in the nonresonance case the depolarization ratios range from 0 to 0.75, in this case values much greater than 0.75 are obtained. This results from the polarizability tensor being unsymmetrical to the point where α_{xy} may equal $-\alpha_{yx}$.

The original work on cytochrome C and hemoglobin has stimulated much work on other proteins and model systems. Only a few representative studies can be cited here. This work has led to detailed discussions of vibronic effects that have not been previously observed. Some of the resonance Raman spectra have been reviewed by Spiro (16) and Carey and Schneider (17).

Fig. 5.10 shows the structure of the synthetic porphyrin, iron tetraphenyl porphine (Fe TPP). This compound is known to form a dimer with a bridging oxygen atom $(Fe(TPP))_2O$. Fig. 5.11 shows the spectra of this compound in CS_2 solution, excited by both 457.9 nm and 571.6 nm radiation (18). Clearly, the relative enhancement of various modes depends on the exciting frequencies. Polarization data allow modes to be separated into three different symmetry classifications. Those with $\rho < 0.75$ belong to A_{1g}, $\rho > 0.75$ are A_{2g}, and $\rho = 0.75$ are B_{1g} or B_{2g} modes.

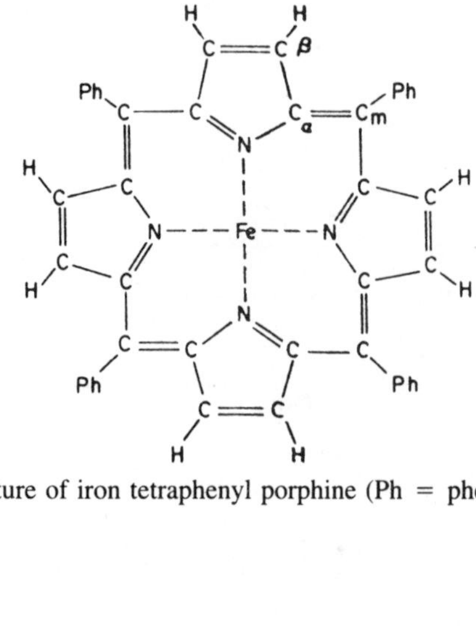

Figure 5.10 Structure of iron tetraphenyl porphine (Ph = phenyl). (From ref. 18.)

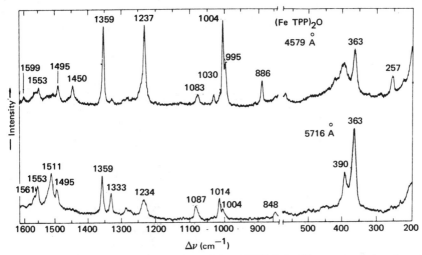

Figure 5.11   Resonance Raman spectra of [Fe (TPP)]₂O in CS₂, concentration 1 mg/ml. (From ref. 18.)

This work provides a good illustration of the analysis of excitation profiles from resonance Raman spectra. These are obtained by tuning a dye laser over the range of the visible spectrum, while measuring the intensity of several Raman bands relative to an internal standard. Results are shown in Fig. 5.12. Strekas et al. (19) have discussed some of the cautions to be observed in measuring such profiles. One sees a variety of Franck-Condon factors at work here, as well as other vibronic mechanisms. Particularly striking is the enhancement observable for the Fe—O—Fe stretch at 363 cm^{-1} (assignment confirmed by isotopic substitution). Spiro and co-workers have discussed how this may occur through interactions with a C—N ring mode at 1359 cm^{-1}.

Spiro and Burke et al. (20) have studied the so-called picket fence porphyrin. This is the only synthetic iron porphyrin currently known to form an isolatable O_2 complex. The Fe—O_2 stretching vibration for this molecule is proven to occur at 568 cm^{-1} (confirmed by removal of O_2 and $^{18}O_2$ substitution). Its frequency is virtually identical with the 567 cm^{-1} observed for oxyhemoglobin. This is a significant finding, since x-ray work has established that the Fe—O—O linkage is bent for picket fence porphyrin. In that case, the identity of the frequencies for Fe—O stretching (and O—O stretching as established from IR spectroscopy) gives excellent assurance that the linkage is bent in hemoglobin as well, confirming Pauling's proposal (21).

Spiro and co-workers have determined the sensitivity of several bands in TPP and picket fence porphyrin to both oxidation state and spin state of the iron. A

87

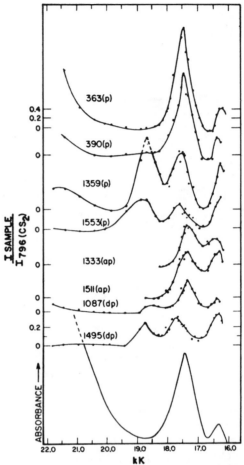

Figure 5.12 Resonance Raman excitation profiles of [Fe (TPP)]$_2$O. The visible absorption spectrum is shown at bottom of figure. Raman intensity is corrected for self-absorption. (From ref. 18.)

summary is shown in Table 5.5. This sensitivity is somewhat different from that found for the naturally occurring hemes. Of particular note is the low value of 1537 cm^{-1}, which occurs in the one five coordinate iron case. This may be diagnostic for five coordination.

Ferris et al. (22) have examined resonance Raman spectra of copper-sulfur complexes. The biological significance here is the potential of these complexes as models for a series of proteins known as blue copper proteins. Another question

Table 5.5   Structure-Sensitive Bands of Iron TPP and $T_{piv}PP$ Complexes[a]

Complex	Frequency (cm^{-1})		
	Band A	Band B	Band C
Fe(III)h.s.[b]	1555	1366	390
Fe(III)ℓ.s.[c]	1568	1370	390
Fe(III)O$_2$[d]	1563	1366	384
Fe(II)i.s.[e]	1565	1370	392
Fe(II)ℓ.s.[f]	1557	1354	382
Fe(II)h.s.[g]	1537	1342	372

From ref. 20.
[a]All three bands are polarized.
[b]High-spin Fe(III): Fe(TPP)Cℓ, Fe(T_{piv}PP)Br, and the μ-oxo dimers.
[c]Low-spin Fe(III): bisimidazole adducts.
[d]Fe(T_{piv}PP) (1-MeIm)O$_2$.
[e]Intermediate spin: four coordinate Fe TPP and Fe T_{piv}PP.
[f]Low-spin Fe(II): bisimidazole adducts.
[g]High-spin Fe(II): 2-methylimidazole adducts.

is just which sulfur ligand is coordinated to copper in these proteins. Four possibilities are shown in Fig. 5.13. By analysis of Cu—S stretching frequencies, excitation profiles, and other aspects of the electronic spectrum, the authors have concluded that proposal 2 (Fig. 5.13), a methionine ligand, is responsible in all of the blue copper proteins that contain methionine.

A rather different use of resonance enhancement was made by Brown et al. (23). The protein lysozyme was among the first studied by laser excited Raman spectroscopy. Brown et al. have excited this spectrum in the UV region (363 nm) to selectively enhance the bands due to tryptophan residues. They have then used this selective enhancement to elucidate the role to tryptophan residues in the binding of lysozyme to glucose. Relative intensity changes are observed in the lysozyme spectrum as a result of this interaction. Again, reference is made to vibronic interactions as they affect the polarizability tensor in order to explain the observed changes.

Bushaw et al. (24) and Gustafson et al. (25) have also used UV radiation (285 to 320 nm) to obtain resonance and preresonance Raman spectra of low solubility proteins and nucleic acid constituents and polynucleotides at micromolar base concentrations.

1. Deprotonated cysteine (mercaptide) coordination:

$$(\text{---})$$
$$\text{protein---S---Cu}^{2+}$$

2. Methionine (thiaether) coordination:

$$\text{protein---S---Cu}^{2+}$$
$$|$$
$$\text{CH}_3$$

3. Protonated cysteine (thiol) coordination:

$$\text{protein---S---Cu}^{2+}$$
$$|$$
$$\text{H}$$

4. Cystine (disulfide) coordination:

$$\text{protein---S---Cu}^{2+}$$
$$|$$
$$\text{protein---S}$$

Figure 5.13   Four possible modes of coordination of sulfur ligands to copper in a Cu-S protein. (From ref. 22.)

The chemistry of vision is known to involve a number of steps, at least one of which is on the picosecond time scale. It has been known for some time that rhodopsin, which consists of a chromophore, all cis retinal, attached to a protein, opsin, isomerizes to all trans retinal and opsin in the photochemical process. This leads to other changes in the protein, which in turn are transmitted to the cell. The state of knowledge on the chemical basis of vision has been reviewed by Honig (26).

Resonance Raman spectra, coupled with various techniques for obtaining spectra in short times, have been used extensively in studying these problems. Early work involved spectra of the model chromophores (27, 28), which concentrated on the effects of protonation of the Schiff base linkage by which the chromophore is attached. This was applied to rhodopsin by several groups (29–31). Normal coordinate calculations have also been reported (32). Lewis and co-workers (33, 34) have taken issue with some of the assignments.

There are a number of intermediates along the photochemical path that had previously been characterized by electronic spectroscopy. These have names such as bathorhodopsin and metarhodopsin. Their spectra are being obtained by time resolved resonance Raman spectroscopic techniques. A number of these techniques have been reviewed (35–37). The combination of flow techniques, laser pulses synchronized with photolyzing pulses, and chopped CW lasers has yielded considerable information on the structure of these intermediates.

Carey and Schneider (17) have utilized the resonance Raman effect in biological systems in yet another way. They use small chromophoric molecules as labels of nonabsorbing biologically active molecules.

The resonance label should be a carefully designed molecule so that its vibrational spectrum can yield the maximum information about the biochemically active site. Essentially, as with naturally occurring chromophores (porphyrin in hemoglobin), only the label spectrum is seen.

This approach has been applied to drug-enzyme interactions, in which the drug is the resonance Raman chromophore, antibody-antigen interaction, and enzyme-substrate reactions. In the enzyme reactions, time dependence or unstable intermediates can sometimes be detected.

An example of the technique is shown in Fig. 5.14. The acyl label is attached to the enzyme chymotrypsin, as shown in the figure. This labeled enzyme is actually an intermediate in the enzyme substrate reaction. The spectrum at pH

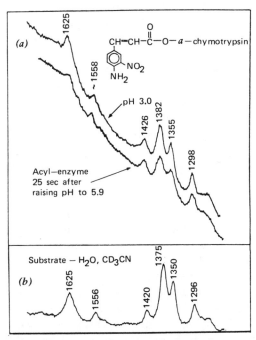

Figure 5.14 Resonance Raman spectra of: *(a)* 4-amino-3-nitro-*trans*-cinnamoyl-α-chymotrypsin at pH 3.0 (top) and pH 5.9 (bottom); *(b)* methyl 4-amino-3-nitro-*trans*-cinnamate. The spectrum of the unstable intermediate at pH 5.9 was obtained in a flow system. Essentially identical spectra were recorded in the range pH 5.9 to 7.0. (From ref. 17.)

91

3 shown in Fig. 5.14 is identical to that of the substrate shown in Fig. 5.14. This enzyme substrate system is stable at pH 3, but when the pH is raised, reaction occurs. Using a flow system, Carey and Schneider obtained the lower spectrum in Fig. 5.14. The major change is a decrease in the intensity of the 1625 cm^{-1} mode, primarily C=C stretching. The authors explain this decrease in terms of a twisting of the

$$
\begin{array}{c}
\diagdown \\
C = O \\
\diagup \\
C = C \\
\diagup
\end{array}
$$

unit out of the plane. Such distortion has often been postulated for substrate reactions but never observed experimentally.

Since the very first work of Spiro and Strekas, it has been clear that resonance Raman spectra of biomolecules requires a detailed knowledge of the basic physics involved in the transitions. As work in this area has progressed, this aspect has not changed. Resonance Raman spectroscopy is not a routine tool for the biologist. Rather it requires a careful spectroscopic analysis. These analyses are solving biological problems while extending our knowledge of vibronic effects.

5.4   LIPIDS AND MEMBRANES

Infrared spectroscopy has been used for some years to study soaps and related lipids. More recently the Raman work has shown great promise for revealing details about membrane structure.

Fig. 5.15 shows a schematic portion of a cell membrane in cross section. Lipid molecules, consisting of polar heads and hydrophobic tails (usually long chain hydrocarbons), assemble in a bilayer in which proteins are immersed. Nutrients must pass through this bilayer into the cell, waste products out.

It is known that lipid-water gels undergo phase transitions as the ratio of lipid to water is changed or as the temperature is increased. It is also known that the principal endothermic phase transition observed on heating is a melting of the hydrocarbon chains. Below this transition the chains are primarily in an all trans, extended conformation; above the transition a significant number of gauche conformers exists. The melting of the hydrocarbon chains greatly affects the fluidity of the membrane, hence the transport properties. Raman spectroscopy has been a good probe of this transition. Indeed, because these are phases of intermediate order and fluidity, Raman spectroscopy is one of the few techniques

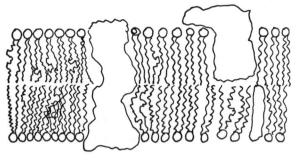

Figure 5.15 Schematic cross section of a biological membrane, showing lipid bilayer with immersed proteins.

that can give information at the molecular level regarding the melting process.

The original Raman spectroscopic work on the phase transition was carried out by Bulkin and Krishnamachari (38) and by Lippert and Peticolas (39). While the sensitivity of the spectrum in the C—H and C—C stretching regions to the trans-gauche ratio in the hydrocarbon chains has been clear for some time, there has been considerable recent discussion over the assignments of the various modes observed and the reasons for their sensitivity to the chain conformation (40, 41).

For dipalmitoyl phosphatidyl choline (dipalmitoyl lecithin), whose structure is shown in Fig. 5.16, there are two thermal phase transitions observed for the

$$
\begin{array}{l}
\quad\quad\quad O \\
\quad\quad\quad \| \\
R^1 —C(O) — C \\
\quad\quad\quad O \quad\quad | \\
\quad\quad\quad \| \\
R^2 —C(O) — C \\
\quad\quad\quad\quad\quad\quad | \quad\quad O^- \\
\quad\quad\quad\quad\quad\quad | \quad\quad | \\
\quad\quad\quad\quad\quad\quad C—O—P—O—CH_2—CH_2—N(R^3)_3^+ \\
\quad\quad\quad\quad\quad\quad\quad\quad \| \\
\quad\quad\quad\quad\quad\quad\quad\quad O
\end{array}
$$

$$R^1, R^2 = C_{15}H_{31}$$
$$R^3 = CH_3$$

Figure 5.16 Structure of lecithin or phosphatidyl choline. For dipalmitoyl phosphatidyl choline the alkyl groups are as indicated.

gels containing 30% or more water. The main transition, already discussed above, occurs at 41°C. At 35°C, however, there is a much weaker endotherm. As shown in Fig. 5.17 using data tabulated from spectra, this transition is also observable by Raman spectroscopy (42, 43). Indeed, the Raman measurements have been interpretable as showing that the transition is a rotation of apposed hydrocarbon chains with respect to one another. This type of phase transition and its thermodynamic properties has many analogues in other solids.

Real biological systems contain mixtures of lipids as well as proteins. There have been a number of attempts to study such systems. Mendelsohn and Maisano (44) showed that if one of the lipids had its hydrocarbon chains deuterated in the mixture, one could easily observe the phase transition of both species in the Raman spectrum. The conformation of each component can thus be monitored. They studied two very similar lipids, differing only in the chain lengths of the hydrocarbon chains. By contrast, work of Bulkin and Krishnamachari (45) on rather dissimilar lipids, in which polar head groups varied, indicated that the behavior of the mixed lipid system was more complicated, involving both eutectic and peritectic phases.

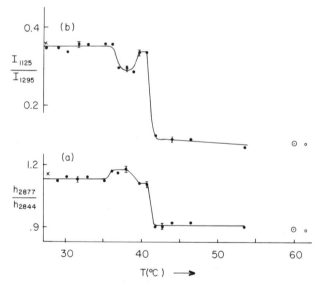

Figure 5.17 Effect of temperature on the Raman spectrum of dipalmitoyl lecithin, 30% water gels: *(a)* relative peak heights in CH stretching region; *(b)* relative peak areas in C—C stretch and CH$_2$ twisting regions. (Unpublished data of Bulkin and Yellin.)

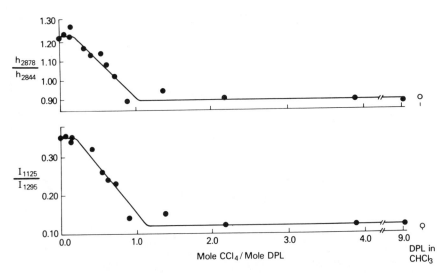

Figure 5.18 Effect on the Raman spectrum of adding carbon tetrachloride to a dipalmitoyl lecithin, 30% water gel: upper, relative peak heights in CH stretching region; lower, relative peak areas in C—C stretch and CH_2 twisting regions. (From ref. 47.)

Lis et al. (46) investigated the interaction of amino acids, simple peptides, and proteins with lecithin. As expected, all the small molecules that they added to the lipid bilayers increased the concentration of gauche conformers. This problem, the effect of small molecules on crystallinity or fluidity, has also been studied by Bulkin and Yellin (47), who investigated the induction of the main phase transition in lecithin-water gels using carbon tetrachloride and chloroform. Fig. 5.18 shows that the fluidity increases over a broad concentration range. In this range there is a two phase region, as predicted thermodynamically for a first-order phase transition. Results such as these may have some importance in elucidating the mechanism of anesthesia.

Membranes and fragments of membranes have been investigated by several groups (48) following the first report of a spectrum of hemoglobin free erythrocyte (red blood cell) ghosts by Bulkin (49) in 1972. Milanovich et al. (50) examined sarcoplasmic reticulum membranes, with a typical spectrum shown in Fig. 5.19. They concluded that this membrane was probably more fluid than the erythrocyte membranes. An important complication in several of the reports is low levels of carotenoids that exhibit resonance enhanced spectra. In the systems we discussed in Section 5.3 under resonance Raman spectroscopy, the presence of the chromophoric group is well known. In biological systems, however, where there

95

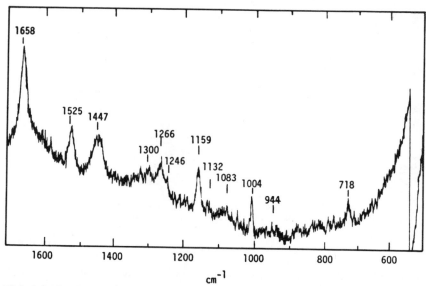

Figure 5.19 Raman spectrum of sarcoplasmic reticulum membranes in H_2O. Maximum signal, 3×10^3 photons/sec; wavelength, 488.0 nm; power, 200 mW; resolution, 4 cm^{-1}; time constant, 30 sec; temperature, 10°C. (From ref. 50.)

is a mixture of many different components, a chromophore may be present at a level of 10^{-3} below that of the species of primary interest, yet interfere strongly with the spectroscopic results.

REFERENCES

1. W. L. Peticolas, *Biochimie* **57**, 417 (1975).
2. R. C. Lord, *Appl. Spectrosc.* **31**, 187 (1977).
3. J. L. Lippert, D. Tyminski, and P. J. Desmeules, *J. Amer. Chem. Soc.* **98**, 7075 (1976).
4. W. S. Craig and B. P. Gaber, *J. Amer. Chem. Soc.* **99**, 4130 (1977).
5. N. T. Yu, T. S. Lin, and A. T. Tu, *J. Biol. Chem.* **250**, 1782 (1975).
6. N. T. Yu, B. H. Jo, and D. C. O'Shea, *Arch. Biochem. Biophys.* **156**, 71 (1973).
7. A. T. Tu, B. H. Jo, and N. T. Yu, *Int. J. Pept. Res.* **8**, 337 (1976).
8. D. B. Boyd, *Int. J. Quantum Chem., Quantum Biol. Sym.* **1**, 13 (1974).
9. G. J. Thomas, *Impact Lasers Spectrosc.* **49**, 127 (1975).

10. K. A. Hartman, R. C. Lord, and G. J. Thomas, Jr., in *Physico-Chemical Properties of Nucleic Acids,* J. Duchesme, Ed., Academic, New York, 1973.

11. G. J. Thomas, Jr., in *Structure and Conformation of Nucleic Acids and Protein-Nucleic Acid Interaction,* M. Sundralingam and S. T. Rao, Eds., University Park Press, Baltimore, 1975, p. 253.

12. L. Lafleur, J. Rice, and G. J. Thomas, Jr., *Biopolymers* **11,** 2423 (1972).

13. M. C. Chan and G. J. Thomas, Jr., *Biopolymers,* 615 (1974).

14. G. J. Thomas and P. Murphy, *Science* **188,** 1205 (1975).

15. T. G. Spiro and T. C. Strekas, *Proc. Nat. Acad. Sci. U.S.A.* **69,** 2622 (1972).

16. T. G. Spiro, *Acc. Chem. Res.* **7,** 339 (1974).

17. P. R. Carey and H. Schneider, *Acc. Chem. Res.* **11,** 122 (1978).

18. J. M. Burke, J. R. Kincaid, and T. G. Spiro, *J. Amer. Chem. Soc.* **100,** 6077 (1978).

19. T. C. Strekas, D. H. Adams, A. Packer, and T. G. Spiro, *Appl. Spectrosc.* **28,** 324 (1974).

20. J. M. Burke, J. R. Kincaid, S. Peters, R. R. Gagne, J. P. Collman, and T. G. Spiro, *J. Amer. Chem. Soc.* **100,** 6083 (1978).

21. L. Pauling, *Nature* (London) **203,** 182 (1964).

22. N. S. Ferris, W. H. Woodruff, D. R. Rorabacher, T. E. Jones, and L. A. Ochrymowycz, *J. Amer. Chem. Soc.* **100,** 5939 (1978).

23. K. G. Brown, E. B. Brown, and W. B. Person, *J. Amer. Chem. Soc.* **99,** 3128 (1977).

24. T. H. Bushaw, F. E. Lytle, and R. S. Tobias, *Appl. Spectrosc.* **32:6,** 585 (1978).

25. T. L. Gustafson, T. H. Bushaw, S. Samanta, R. S. Tobias, and F. E. Lytle, *Proc. 7th Int. Conf. Raman Spectrosc.,* North-Holland, Amsterdam, 1980, p. 588.

26. B. Honig, *Ann. Rev. Phys. Chem.* **29,** 31 (1978).

27. L. Rimai, D. Gill, and J. Parsons, *J. Amer. Chem. Soc.* **93,** 1353 (1971).

28. M. Heyde, D. Gill, R. Kilponen, and L. Rimai, *J. Amer. Chem. Soc.* **93,** 6776 (1971).

29. R. H. Callender, A. Doukas, R. Crouch, and K. Nakanishi, *Biochemistry* **15,** 1621 (1976).

30. A. R. Oseroff and R. H. Callender, *Biochemistry* **13,** 4243 (1974).

31. R. Mathies, A. R. Oseroff, and L. Streyer, *Proc. Nat. Acad. Sci. U.S.A.* **73,** 1 (1976).

32. B. Aton, R. H. Callender, and B. Honig, *Nature* (London) **273,** 784 (1978).

33. M. Sulkes, A. Lewis, and M. A. Marcus, *Biochemistry* **17,** 4712 (1978).

34. A. Lewis, *Proc. Nat. Acad. Sci. U.S.A.* **75,** 549 (1978).

35. M. A. El-Sayed. *Proc. VIIth Int. Conf. Raman Spectrosc.,* W. F. Murphy, Ed., North-Holland, New York, 1980, p. 542.

36. R. Mathies, G. Eyring, B. Curry, A. Broek, I. Palings, R. Fransen, and J. Lugtenberg, *Proc. VIIth Int. Conf. Raman Spectrosc.*, W. F. Murphy, Ed., North-Holland, New York, 1980, p. 546.

37. R. Callender, *Proc. VIIth Int. Conf. Raman Spectrosc.*, W. F. Murphy, Ed., North-Holland, New York, 1980, p. 548.

38. B. J. Bulkin and N. Krishnamachari, *J. Amer. Chem. Soc.* **94,** 1109 (1971).

39. J. L. Lippert and W. L. Peticolas, *Proc. Nat. Acad. Sci. U.S.A.* **68,** 1572 (1971).

40. B. P. Gaber, P. Yager, and W. L. Peticolas, *Biophys. J.* **44,** 21, 161 (1978).

41. B. P. Gaber, P. Yager, and W. L. Peticolas, *Biophys. J.* **44,** 677 (1978).

42. B. P. Gaber and W. L. Peticolas, *Biochim. Biophys. Acta* **465,** 260 (1977).

43. B. J. Bulkin and N. Yellin, unpublished data.

44. R. Mendelsohn and J. Maisano, *Biochim. Biophys. Acta* **506,** 192 (1978).

45. B. J. Bulkin and N. Krishnamachari, *Mol. Cryst. Liq. Cryst.* **24,** 53 (1973).

46. L. J. Lis, J. W. Kauffman, and D. F. Shriver, *Biochim. Biophys. Acta* **436,** 513 (1976).

47. B. J. Bulkin and N. Yellin, *J. Phys. Chem.* **82,** 821 (1978).

48. J. L. Lippert, L. E. Gorczyka, and G. Meikeljohn, *Biochim. Biophys. Acta* **382,** 51 (1975).

49. B. J. Bulkin, *Biochim. Biophys. Acta* **274,** 649 (1972).

50. F. P. Milanovich, Y. Yeh, R. J. Baskin, and R. C. Harney, *Biochim. Biophys. Acta* **419,** 243, (1976).

Chapter Six

Liquid Crystals

Infrared and Raman spectroscopy have been used for many years to study strong intermolecular interactions such as hydrogen bonding. It is natural that many research groups should have thought to apply these techniques to the study of liquid crystals.

Although studies of the vibrational spectra of liquid crystals began many years ago, there has been a flurry of activity in the years since 1968. Work in this area has been the subject of review (1).

The application of vibrational spectroscopy to mesophases has proceeded in stages. In the early work, and in some of the work still being published today, the main contribution has been the observation of phenomena in the spectra—intensity changes, frequency shifts, and so on—that occur at or near the various phase transitions. Further, as is characteristic of the application of spectroscopic studies to liquid crystals, a certain amount of attention has been focused on measurement of order parameters.

There now seems to be sufficient background available, however, to begin to make some generalizations about the expectations one has for the vibrational spectra of liquid crystals. Further, one can begin to make some quantitative assignment of the phenomena observed to changes at the molecular level.

Liquid crystals are ordered, fluid phases that occur for several thousand organic compounds and polymers between the crystalline and isotropic liquid phase. Such phases are also present in mixtures of compounds. For certain of the latter cases, the individual compounds may not in themselves form liquid crystals, while the mixture, at certain concentrations and temperatures, does. These are known as lyotropic liquid crystals, while materials that, when pure, show liquid crystalline behavior, are known as thermotropic liquid crystals. Metastable liquid crystalline phases also exist, that is, materials exist that show simple phase

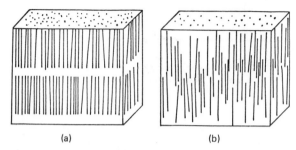

Figure 6.1 Schematic view of molecular arrangements in: *(a)* the smectic; and *(b)* the nematic liquid crystalline states. There are other smectic phases as well, for example, those in which the molecules tilt in the layers.

transitions from crystal (c) to isotropic liquid (ℓ) on heating, but that exhibit a liquid crystalline phase on cooling. These are known as monotropic liquid crystals.

Liquid crystals can be divided into several classes, according to the type of long range order that is present. In general, crystalline phases are characterized by three-dimensional order, whereas liquid crystalline phases possess at most two-dimensional order. Fig. 6.1*a* shows a schematic diagram of the order of molecules in a smectic liquid crystal. Here the molecules are arranged in layers, with their long axes parallel within the layers. No other order exists within the layers, however. The figure also introduces the idea that the molecules must be anisotropic in shape for liquid crystalline behavior to be found.

Fig. 6.1*b* shows a different type of liquid crystal, the nematic phase. In this phase. the long axes of the molecules are parallel, but no other order exists. It is perhaps not surprising that some molecules, on heating, show first a smectic and then a nematic phase before being converted to the isotropic liquid. A third general class of liquid crystals are cholesteric liquid crystals, in which the molecules are arranged in two-dimensional layers that look like a cut through a nematic phase, but successive layers are twisted with respect to one another, so that if one were to follow the direction of the long axes of the molecules through successive layers it would describe a helix. While some work has been done on the vibrational spectra of cholesteric phases (2), it is not discussed in this chapter.

6.1 CHAIN FLUIDITY

The question of chain fluidity in nematic and smectic phases has been the subject of several Raman investigations (3). Schnur (4) has studied the alkoxy azoxy benzenes (Fig. 6.2) in all spectral regions, but with particular emphasis on the

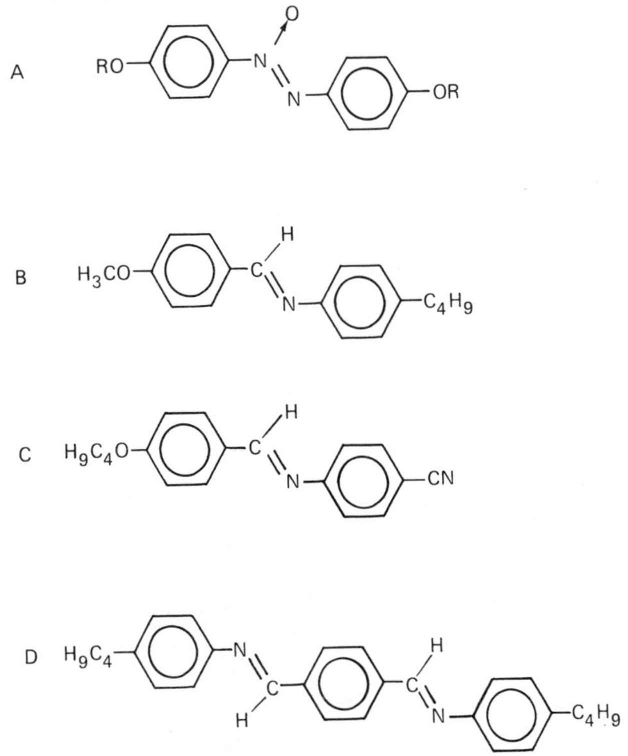

Figure 6.2 Structures of some molecules that form liquid crystals: *(a)* The alkoxy-azoxybenzene series; *(b)* 4-methoxybenzylidene-4'-*n*-butylaniline (MBBA); *(c)* 4-meth-oxybenzylidene-4'-cyanoaniline (BBCA); *(d)* 4,4'-bisterephthalbisbutyl aniline (TBBA).

region from 200 to 400 cm^{-1}, where it is asserted that an accordion-like vibration of a short (3 to 8 carbon) all trans hydrocarbon chain attached to a benzene ring should occur. An approximate calculation (3) was carried out to confirm the assignment, although again one must assume that there is some mixing with ring modes and other modes of the chain.

The results of these studies indicate that near the crystal-nematic (c-n) transition, and to some extent in the nematic phase as well, there are conformational changes taking place in the end chains. An illustration of the changes observed for the C_6 homolog is shown in Fig. 6.3. Of course, these compounds also show other spectral changes as discussed above, but the changes observed by Schnur seem to have a distinct phase and temperature dependence. In some of the compounds, solid phase polymorphism exists, and this seems to drastically affect the intensity of the accordion mode. An example for the C_4 homolog is shown

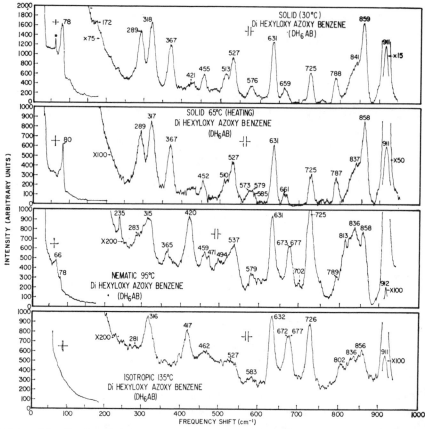

Figure 6.3 Tracing of observed Raman spectra of dihexyloxyazoxybenzene (C₆). (From ref. 4.)

in Fig. 6.4. Note that this might be interpretable as a pretransition effect in the crystal rather than a distinct solid phase.

Additional evidence regarding chain fluidity comes from a paper by Bulkin et al. (5), who showed that in the nematic phase of N-(4-methoxybenzylidene)-4'-butylaniline (MBBA) the CH stretching vibrations show an increase in band width, so that the entire contour of the CH stretching modes appear to lose resolution. As temperature is increased until just below the nematic-isotropic phase transition, the bands sharpen with a slight shift in the position of the frequency maxima. This behavior was interpreted as arising from a reorganization of butyl chain conformations on the time scale of the IR measurement. Since the two time scales (internal rotation and CH stretching) overlap for a short temperature range, broadening is observed. Such phenomena are common in NMR. Similar reorganization of conformers in MBBA was deduced from ultrasonic measurement of Jahnig (6) and is consistent with these data.

Lugomer and Lavrencic (7) have reexamined the Raman intensity data for a wide variety of nematic materials with different length carbon end chains in an attempt to interpret these data in terms of the elastic constant ratio K_{33}/K_{11}, which shows a very strange temperature dependence in the nematic phase (8).

At the nematic to isotropic (n-ℓ) transition, there are few, if any, real changes observed in the spectra. One likely candidate is the change in intensity of the accordion mode observed by Schnur (3). There has been confusion on this transition from Raman spectroscopic observations in the past.

This confusion arises as follows: In the Raman spectrum of a low symmetry, nematogenic molecule, the Raman bands are all polarized but may have widely varying depolarization ratios. In an unaligned or imperfectly aligned sample, the Raman exciting radiation and the scattered light have their electric vectors rotated due to refractive index discontinuities. Such an effect is well-known in the Raman spectrum of powders (9). When the n-ℓ transition occurs, this scrambling of polarization disappears, and the result is a different set of relative intensities.

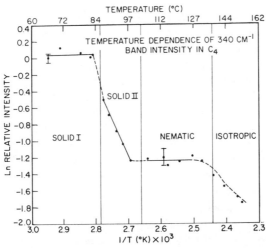

Figure 6.4 Relative integrated intensity in logarithmic units of the accordion band in C_4 versus reciprocal temperature. The integrated intensity of the 340 cm^{-1} band was calculated relative to the adjacent 320 cm^{-1} band, which appeared to remain reasonably constant. (From ref. 4.)

103

This has been documented in a more quantitative fashion by Bulkin et al. (10). Thus any changes in relative intensity of bands (and changes in apparent frequency maxima that may occur from overlapping bands changing in relative intensity) must be carefully checked by study of the IR spectra in most cases.

6.2 POLARIZATION AND ORDER PARAMETERS

The question of measuring polarization in the Raman spectrum of liquid crystals has been a difficult one. Some initial attempts were unsuccessful, with authors observing a depolarization ratio of 1.0 in the nematic phase. This is indicative of complete scrambling of the radiation. If changes from a ratio of 1.0 are observed at the n-ℓ transition, as has been reported (11), then this is not surprising.

The most thorough treatment of oriented nematics in the Raman spectrum has been done by Pershan and co-workers (12, 13). They have shown that it is, in principle, possible to measure both the first $<\cos^2 \theta>$ and second $<\cos^4 \theta>$ terms in the order parameter or distribution function from the Raman spectrum. Experimentally, they have used samples of MBBA in which a probe molecule that also forms a nematic phase *N-p*-butoxybenzyliden-*p'*-cyanoaniline (BBCA) is dissolved. (See Fig. 6.2 for structures of these molecules.) The sample is aligned by rubbing two plates, and 180° Raman scattering is observed. The BBCA molecule shows a strong CN stretching vibration, which is approximately along the long axis and is isolated from other intramolecular vibrations. From these measurements, the terms in the order parameter can be deduced. Comparison with existing mean field theories of the nematic order parameter has been made from these measurements. It is asserted that the fit is better with the Humphries et al. (14) approach than with the Maier-Saupe (15) theory for the $<\cos^2 \theta>$ term, but that neither does particularly well with the $<\cos^4 \theta>$ observations. The authors have discussed possible reasons for the strange $<\cos^4 \theta>$ results, including those that may arise from the measurement technique itself.

Boyd and Wang (16) have applied the Raman technique to the study of the effect of pressure on nematic materials. In this case, the Raman spectrum simply becomes the analytical probe by which the phase transition is detected. It was shown that the c—n and n—ℓ transitions are effectively first-order (follow the Clausius-Clapeyron equation) when measured in this way for one case. This technique is useful in that it also allows one to determine the change in volume at the phase transition; in both cases (c—n and n—ℓ) the higher temperature phase had the lower density.

6.3 EXTERNAL MODES

One of the areas in which vibrational spectroscopy yields quite unique information is the study of intermolecular or lattice vibrations. These are seen close to the exciting line in the Raman spectrum; hence a good rejection of stray light and a relatively low Rayleigh scattering level are needed for their observation. In the IR spectrum, the lattice modes occur in the far infrared region, generally below 150 cm^{-1} for organic crystals. They tend to absorb very strongly, particularly in molecules such as nematogenic crystals, which usually have dipolar groups present. Whereas in the Raman, spectrum measurements with single crystals are readily made and yield definitive assignment of the lattice modes to a particular symmetry species, in the far infrared single crystal measurements are difficult. This further complicates the observations, because the crystallite size may be comparable to the wavelengths of the radiation, leading to spectral distortions.

In the nematic phase we have the possibility of observing "pseudo-lattice" vibrations, that is, intermolecular motions characteristic of the long range order in this phase. It should be noted that the intermolecular potential in such a phase with organic crystals probably is proportional to an intermolecular separation term of $1/r^3$ or higher power of $1/r^n$, so the force constants, which depend on the second derivative of $1/r^n$, fall off rather rapidly with distance. However, in cases where there is long range order, it may be possible to propagate phonons even in liquid crystals. As we see, there is not much evidence for this in the spectra of nematics, but there is some in the case of the more highly ordered smectic phases.

Although a substantial number of papers on nematics have appeared, almost all the attention has been concentrated on two materials, *p*-azoxyanisole (PAA) and *N*-(*p*-methoxybenzylidene)-*p*-butylaniline (MBBA) (Fig. 6.2). We discuss these two cases in some detail, referring relatively briefly to other work that has been carried out.

The Raman spectrum of PAA in the lattice vibration region was first published by Bulkin and Grunbaum (17). Bulkin and Prochaska (18) examined the Raman spectrum of a single crystal of PAA at −90°C, and were thus able to assign the lattice vibrations to particular symmetry species. Their spectrum, taken in two views to show the A_g and B_g modes clearly, is shown in Fig. 6.5.

As the c-n transition is approached, Bulkin and Prochaska (18) observed pretransition effects that they interpreted as being indicative of a soft mode, such as is commonly observed near ferroelectric phase transitions. They asserted that this soft-mode-like behavior occurs with a mode or modes that initially (i.e., 5° below the phase transition) are in the vicinity of 70 cm^{-1}. It appears to shift

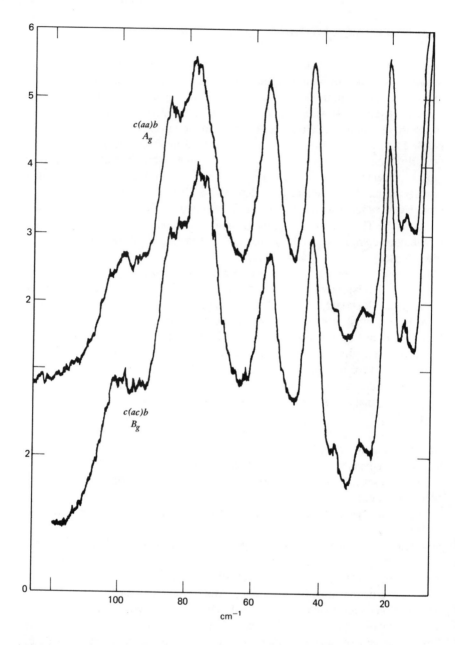

Figure 6.5 Raman spectra of *p*-azoxyanisole (PAA) at −90°C showing the A_g and B_g modes.

rapidly toward zero frequency in the 2 to 3° below the phase transition. Amer and Shen (19) did not find evidence for such pretransition effects and attributed this observation to problems with temperature control in the experiments of Bulkin and Prochaska. Sakamoto et al. (20) also failed to observe this effect, but in their case the lattice vibration spectrum even 8° below the phase transition shows only two broad bands, instead of six distinct maxima observed by other workers.

There is other evidence for the pretransition effects described above. First, it should be pointed out that to a certain extent there is a matter of definition of a pretransition effect. Some workers have seen the transition occurring over a 2 to 3° range and stated that it was abrupt, yet this is the range over which it was asserted that the soft mode behavior occurred. Secondly, Riste and Pynn (21), using neutron scattering techniques, have observed identical pretransitional behavior at the PAA phase transition. As we see below, there are pretransition effects observed in the Raman spectra of other liquid crystalline materials at the c-n phase transition. The soft mode behavior was predicted in theoretical treatments of Kobayashi (22) and Ford (23).

Bulkin and Grunbaum (24) have carried out a calculation of the lattice vibrations of PAA. In Table 6.1 the results of their calculation are shown, compared with experimental data from Bulkin and Prochaska (18) and Bulkin and Lok (25). The calculation is based on a model potential function for the unit cell, which uses atom-atom interactions of the form

$$V_n = \frac{-a}{r^6} + be^{-cr}$$

where r is the nth atom-atom distance and a, b, and c are parameters. In the calculation the parameters are transferred from the literature without any refinement on the experimental data. The potential is differentiated twice with respect to all motional coordinates to give the force constants, from which the frequencies are then calculated.

The results in Table 6.1 are grouped according to motions about the same type of coordinate. Thus when the frequencies in a particular column are close to one another, this is indicative of weak coupling between the molecules in the unit cell with respect to motion along that coordinate. A large splitting reflects a strong coupling. As expected for anisotropic molecules, both cases occur. It is interesting to note that the lowest and highest frequency bands observed in the spectrum result from very strong coupling, giving rise to a large splitting. Likewise, the case of weak coupling leads to bands in the 70 cm^{-1} region, which is where the putative soft mode was seen. This is as expected for such a mode. This calculation seems possible now for all nematogenic or smectogenic crystals

Table 6.1 Observed[a] and Calculated Lattice Vibration Frequencies of PAA

Symmetry Species	Translatory[b]			Rotatory[b]		
A_g						
Observed	30	52	70	16	74	91
Calculated	28	55	62	20	69	94
B_g						
Observed	30, 37	52	70	16	74, 90	95
Calculated	37	60	69	25	82	101
A_u						
Observed	50	70	—	135, 150	84	50?
Calculated	51	69	0[c]	130	89	55
B_u						
Observed	—	—	70?	115	84?	50?
Calculated	0[c]	0[c]	67	119	78	52

From ref. 24.

[a]Raman data (A_g, B_g) from ref. 18, single crystal frequencies at $-90°$. Far infrared (A_u, B_u) data from ref. 25, polycrystalline sample at 25°.
[b]Modes are primarily rotatory or translatory in nature; however, the translation-rotation interaction force constants are not negligible.
[c]Acoustical modes.

for which crystal structure data are available. Raman spectra of the higher homologs of PAA have been studied by several groups in the lattice vibration region. Interestingly, none shows as well-defined a spectrum as does PAA. Bulkin et al. (26) had previously commented that it seemed, on the basis of IR data, to be more difficult to obtain highly ordered crystals of these materials. Schnur (4) has pointed out that there is considerable solid polymorphism in these materials, and in macroscopic samples one wonders whether this might be affecting the resolution in the lattice vibration spectrum.

What of the spectra in the nematic phase? In all papers, the authors seem to agree that there is Raman scattering in the region below 100 cm^{-1}, which appears in the nematic phase. There are distinct frequency maxima on the Rayleigh wing, which subsequently disappear in the isotropic phase. Not surprisingly, there are differences in the positions of the quoted frequency maxima. These are due to the difficulty of accurately measuring the positions of broad bands on an intense background. It would seem desirable to carry out some computer deconvolution of the Rayleigh background in an attempt to find these bands more accurately. This information will be valuable in any attempt to model the nematic inter-molecular vibrations. Also somewhat controversial is the question of whether the observed intermolecular Raman bands in the nematic phase disappear con-tinuously or discontinuously at the n-ℓ phase transition.

The other major class of compounds that has been studied in the low frequency region is the Schiff base nematics. MBBA has been studied in the most detail. The Raman spectrum of MBBA in the lattice vibration region was first reported by Billard et al. (27), and subsequently by Borer et al. (28) and Vergoten et al. (29). The spectrum of Billard et al. is shown in Fig. 6.6 at two temperatures in the solid state. At low temperatures, even in the polycrystalline sample, MBBA shows a very well-defined vibrational spectrum. Nine distinct frequency maxima are observed between 39 and 180 cm^{-1} at $-173°C$, but the two highest modes,

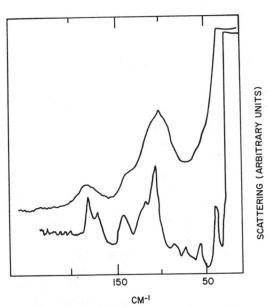

Figure 6.6 Raman spectrum of solid MBBA in the lattice vibration region: upper curve, 15°C; lower curve, 100°C. (From ref. 27.)

at 179 and 181, are explained as internal vibrations. It seems possible that the mode at 142 cm^{-1} may also have a contribution from internal vibrations; this could be checked by study of solution spectra. Borer et al. are in essential agreement with these results.

At temperatures close to the c-n phase transition, Billard et al. have noted that the spectrum becomes quite broad and most of the fine detail is lost. Thus in the spectrum at 15°C, about 5 to 7° below the c—n phase transition, the bands are already quite broad. A spectrum given by Borer et al. at 2°C also shows considerable broadening. This seems to be indicative of pretransition effects similar to those described earlier, although no soft mode was observed.

6.4 SMECTIC PHASES

Some work has been done on the Raman spectra of thermotropic smectic phases. Amer and Shen (30) studied the Raman spectrum of diethylazoxybenzoate and diethylazoxycinnamate, which have smectic A phases. These compounds show a single low frequency Raman mode in the crystalline phase at 22 and 26 cm^{-1}, respectively. In the smectic phase, this mode appears to shift, with the maximum at about 14 cm^{-1}. It vanishes abruptly at the smA-ℓ transition. The mode is strongly overlapped by the Rayleigh line in the smectic phase, but it does seem to be of considerably lower intensity than that seen by neutron scattering discussed above. As usual, the modes are explained as being some sort of rotatory oscillations, but in this case one that is primarily associated with strongly interacting functional groups, such as C=O, within the molecule. This would account for the seemingly small mass dependence in the smectic phase. Probably such an explanation is not necessary, since a relatively small frequency shift in such a low frequency mode would be sufficient to account for the mass difference. The authors have also studied the internal vibrations of this compound, as have Zhdanova et al. (31). No frequency shifts or relative intensity changes with temperature, such as have been seen with nematic materials, were observed.

The interesting compound terephthalbisbutyl aniline (TBBA) (Fig. 6.2) has been studied. This material has seven known fluid phases, namely, isotropic, nematic, three stable smectics (B or H, C, and A), and two monotropic liquid crystalline phases not yet characterized.

Schnur and Fontana (32) studied all of the stable phases in the low frequency and higher frequency Raman spectra. They have two important sets of observations. In the low frequency region, the lattice vibration spectrum, which contains a number of bands, appears in their spectra to be almost the same in the smectic B phase as it is in the crystal. At the smB-smC transition, most of this structure disappears, and the spectra in the higher temperature phase all show only the broad scattering characteristic of these phases. Note that a band is observed by these workers in the 22 cm^{-1} region of the crystal, which shifts to 19 cm^{-1} in the smB phase and disappears in the higher temperature phases. If this were the same mode occurring in this region referred to earlier, we would expect to see it persist in the higher temperature phases.

A curious observation is seen in the 1550 to 1650 cm^{-1} region. In the solid and smB phases, one band is seen in this region at 1590 cm^{-1}. This is also true in the smB phase, but in the smC phase two new bands appear, at about 1560 and 1620 cm^{-1}. This has not been observed in any other spectra of liquid crystals. The authors do not expound on the possible explanations of this phenomenon. Nonetheless, mechanisms can be enumerated by which new bands could appear in the spectrum. In particular, the most usual are the appearance of new conformations (in this case perhaps due to the multiple minimum potential that likely exists about the Schiff base linkages), or a breaking of symmetry, resulting in a lowering of selection rules. In this regard, it should be noted that the smB phase is expected to have a higher symmetry than the smC phase. In addition, TBBA is one of the few liquid crystalline molecules that has a fairly high molecular symmetry; it probably possesses an approximate or exact center of symmetry. The intersection of the molecular symmetry and smB symmetry could result in selection rules that would cause modes to be forbidden in smB but allowed in lower symmetry phases. This point should be explored further, as other molecular modes should show the same behavior if this is the case.

Dvorjetski et al. (33) have also investigated TBBA, concentrating their attention on the low frequency region Raman spectrum in the lower temperature phases (smB and the two monotropic phases, called VI and VII). In the solid phase their observations are in agreement with those of Schnur and Fontana, but this is not true in the smB phase. Here at 130°C (the same temperature that Schnur and Fontana used), Dvorjetski et al. only observed broad, ill-defined scattering. In phase VI, at 80°C, they observed some additional structure, while in phase VII, at 67°C, the original crystalline spectrum is recovered intact. There is no apparent explanation for the difference in observations between these two groups, and resolution must await further experimental work.

Dvorjetski et al. have studied the temperature dependence of the 19 cm^{-1} band in the solid state, approaching the c-smB transition. After correcting for background and Boltzmann factor, they find that this mode shows a small pre-transition effect in its frequency, shifting toward zero frequency over a temperature range of about 4°C. They thus feel that this mode behaves as a soft mode and interpret their results in terms of the types of molecular motion possible in the various phases. They conclude that the data are consistent with free molecular rotation about the long axis in the smB phase. This rotation freezes out progressively as the molecules are cooled to phases VI and VII. Further work is needed to verify these conclusions in other regions of the spectrum.

REFERENCES

1. B. J. Bulkin, *Adv. Liq. Cryst.* **2,** 199 (1976).
2. B. J. Bulkin and K. Krishnan, *J. Amer. Chem. Soc.* **93,** 5998 (1971).
3. J. M. Schnur, *Phys. Rev. Lett.* **29,** 1141 (1972).
4. J. M. Schnur, *Mol. Cryst. Liq. Cryst.* **23,** 155 (1973).
5. B. J. Bulkin, D. Grunbaum, T. Kennelly, and W. B. Lok, *Liq. Cryst., Proc. Int. Conf. 1973, Pramana Suppl. No. 1,* p. 155.
6. F. Jahnig, *Chem. Phys. Lett.* **23,** 262 (1973).
7. S. Lugomer and B. Lavrencic, *Solid State Comm.* **15,** 177 (1974).
8. H. Gruler, *Z. Naturforsch, Teil A* **28,** 474 (1972).
9. B. J. Bulkin, *J. Opt. Soc. Amer.* **59,** 1387 (1969).
10. B. J. Bulkin, J. O. Lephardt, and K. Krishnan, *Mol. Cryst. Liq. Cryst.* **19,** 295 (1973).
11. C. H. Wang, *J. Amer. Chem. Soc.* **94,** 8605 (1972).
12. E. B. Priestly, P. S. Pershan, R. B. Meyer, and D. H. Dolphin, *Vijnana Parishad Anusandhan Patrika* **14,** 93 (1971).
13. E. B. Priestly and P. S. Pershan, *Mol. Cryst. Liq. Cryst.* **23,** 369 (1973).
14. R. L. Humphries, P. G. James, and G. R. Luckhurst, *J. Chem. Soc., Faraday Trans. 2* **68,** 1031 (1972).
15. W. Maier and A. Saupe, *Z. Naturforsch, Teil A* **15,** 287 (1960).
16. J. D. Boyd and C. H. Wang, *J. Chem. Phys.* **60,** 1185 (1974).
17. B. J. Bulkin and D. Grunbaum, in *Liquid Crystals and Ordered Fluids,* Vol. 1, R. Porter and J. Johnson, Eds., Plenum, New York, 1970.
18. B. J. Bulkin and F. T. Prochaska, *J. Chem. Phys.* **54,** 635 (1971).
19. N. M. Amer and Y. R. Shen, *J. Chem. Phys.* **56,** 2654 (1972).
20. A. Sakamoto, K. Yoshino, U. Kubo, and Y. Inuishi, *J. Appl. Phys.* **13,** 1691 (1974).
21. T. Riste and R. Pynn, *Solid State Comm.* **12,** 407 (1973).
22. K. K. Kobayashi, *Mol. Cryst. Liq. Cryst.* **13,** 137 (1971).
23. W. G. F. Ford, *J. Chem. Phys.* **56,** 6270 (1972).
24. B. J. Bulkin and D. Grunbaum, *J. Phys. Chem.* **79,** 821 (1973).
25. B. J. Bulkin and W. B. Lok, *J. Phys. Chem.* **77,** 326 (1973).
26. B. J. Bulkin, D. Grunbaum, and A. Santoro, *J. Chem. Phys.* **51,** 1602 (1969).

27. J. Billard, M. Delhaye, J. C. Merlin, and G. Vergoten, *C. R. Acad. Sci., Ser. B* **273,** 1105 (1971).

28. W. J. Borer, S. S. Mitra, and C. W. Brown, *Phys. Rev. Lett.* **27,** 379 (1971).

29. G. Vergoten, R. Demal, and G. Fleury, *Trav. Soc. Pharm. Montpellier* **33,** 321 (1973).

30. N. M. Amer and Y. R. Shen, *Solid State Comm.* **12,** 263 (1972).

31. A. S. Zhdanova, L. F. Morozova, G. Peregudov, and M. Sushchinskii, *Opt. Spektrosk.* **26,** 209 (1969).

32. J. M. Schnur and M. Fontana, *J. Phys. (Paris) Lett.* **35,** 53 (1974).

33. D. Dvorjetski, V. Volterra, and E. Wiener-Avnear, *Phys. Rev. A* **12,** 681 (1975).

Chapter Seven

Inorganic and Organometallic Chemistry

The importance of Raman spectroscopy in the field of inorganic and organometallic chemistry is evident by the large number of books, review chapters, and articles that are available (1–9). It is not only that Raman data are essential for the complete structural analysis of molecules, but it is again the ease of obtaining Raman spectra from colored solids, in the molten state, or in solution, that has contributed to the recent growth in this field. The review by Cody and Darlington (9), emphasizing practical analytical applications of Raman spectroscopy, includes characterization of polymorphism in antimony oxides, quantification of minor impurities in metal oxides, and solid state phase transitions as a function of temperature. The enhancement of intensity of Raman bands when resonance Raman conditions are met has generated an increase in activity in this area. Excellent reviews by Shorygin (10) and Clark and Stewart (11) summarize the field.

7.1 STRUCTURAL STUDIES

7.1.1 Molecular Species

The vibrational analysis of crystalline inorganics to assign internal modes of the ions, as well as crystal lattice modes, requires both IR and Raman spectroscopy. Many examples of such structural studies are found in texts and review articles.

Baran (12) measured the IR and Raman spectra of crystalline $MgTe_2O_5$ and assigned the internal vibrations of the $Te_2O_5^{2-}$ ion. He also did structural studies on crystalline magnesium orthovanadate (13). Infrared and Raman spectra of $BaCl_2 \cdot 2H_2O$ powder were obtained and interpreted by Jain et al. (14). Crystalline Na_2CrO_4 and $Na_2CrO_4 \cdot 4H_2O$ were compared by Carter and Bricker (15), and the effects of hydrogen bonding on the chromate frequencies noted. Carter and O'Hare subsequently studied $(NH_4)_2CrO_4$ and $(ND_4)_2CrO_4$ in a special rotating cell designed for liquids but modified for solid samples (16).

The utility of Raman spectroscopy in structural investigations of systems containing transition metal to transition metal bonds was shown by San Filippo and Sniadoch (17). They found that the Raman spectra of simple homonuclear transition metal complexes showed an intense, sharp band in the low frequency region due to the metal-metal stretching vibration. These metal-metal frequencies showed little variation with respect to the nature of the coordinated ligands but were considerably dependent on the order of the metal-metal bond.

Another example of Raman structural studies involves the polymorphs of cobalt molybdate (18). The tetrahedral, purple, high temperature form (β) is metastable at room temperature and converts to the octahedral, green, low temperature form (α) with any pressure application. Therefore, it is impossible to prepare any of the α form for IR examination. Raman spectroscopy can be used, and Angell first obtained the spectra of both the α and β forms in 1972 (19).

Numerous other cobalt, molybdenum, tungsten, and rare earth compounds have been examined by Raman spectroscopy. Py et al. (20) studied molybdenum trioxide, MoO_3, and suggested that the coordination of the oxygen atoms around the molybdenum is tetrahedral rather than octahedral. Miller obtained the Raman spectrum of a single crystal of magnesium molybdate (21), but because of the complexity could not meaningfully assign bands to tetrahedral modes despite the fact that the oxygens were tetrahedrally coordinated about the molybdenum atoms. Knözinger and Jeziorowski have studied a series of molybdenum oxide catalysts supported on aluminas and have found polymeric aggregates of MoO octahedra on the surface (22).

A series of vibrational spectroscopic studies of molybdates, tungstates, and related compounds were done by Liegeois-Duyckaerts. He obtained the IR and Raman spectra of the three hexagonal perovskites $Ba_2B^{II}TeO_6$ (B^{II} = Ni, Co, Zn) (23) and compared them with those of the corresponding cubic perovskites (24). Based on this comparison, he made a general assignment of the internal modes and discussed in detail two A_{1_g} modes that he assigned to the symmetric stretch of two different types of TeO_6 octahedra.

Isotopic substitution and group analysis aided in the IR and Raman structural studies of M_2TiO_5 compounds (M = rare earths: La to Dy). Paques-Ledent (25) found one interesting experimental feature occurring simultaneously in the IR and Raman spectra of these compounds—a band in the frequency range 775 to 875 cm^{-1}, which is an exceptionally high value and signifies one short distance

115

Ti—O bond in M_2TiO_5 compounds. This is particularly interesting because the fifth oxygen of a TiO_5 polyhedron is structurally different from the other four oxygen atoms, which are strongly bonded to double groups M_IO_7—$M_{II}O_7$. Therefore, the comparison between the crystallographic study and the vibrational study allowed the detection of a local vibration in an inorganic crystal. Rao (26) has reviewed the Raman spectroscopic studies of complex metal oxides.

7.1.2 Ionic Species

There has been very extensive work, primarily by physicists, on the Raman spectra of ionic single crystals. This work has applicability to understanding the electrical and electrooptical properties of these crystals. An excellent bibliography of work up to 1974 is available (27); it contains 585 references to individual studies of crystals.

The use of a lasar source for Raman spectroscopy greatly increased its usefulness for structural determinations of highly colored samples. Even though the compounds were colored, the Raman spectra of molecular metal halides, MX_4 (M = C, Si, Ge, Ti, or Sn; X = F, Cl, Br, or I) have been obtained (3). All four fundamentals expected for molecules of T_d symmetry—$v_1(a_1)$, $v_2(e)$, $v_3(t_2)$, and $v_4(t_2)$ (only the t_2 fundamentals are IR active)—were observed. Fig. 7.1 shows the spectrum of vanadium oxytribromide, which is dark red in color (28).

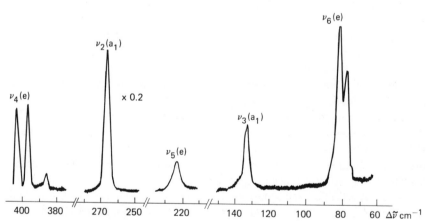

Figure 7.1 Raman spectrum of crystalline vanadium oxytribromide, 647.1 nm excitation. Symmetry designations are given in C_{3v} nomenclature. (From R. J. H. Clark and P. D. Mitchell, *J. Chem. Soc., Dalton*, 2429 (1972).)

The appearance of six fundamentals (three of which are polarized) is consistent with C_{3v} symmetry.

7.1.3   Colored Coordination Compounds

Many coordination compounds are deeply colored. Work has been undertaken to obtain Raman values of fundamentals that, when combined with IR data, can be used for calculations of force constants and bond strengths. Examples include studies of the square planar platinum or palladium complexes of the type $MX_2(SR_2)_2$ (X = Cl, Br, or I; R = Me, Et, isopropyl, etc.) (28, 29). The biggest hindrance to the study of these colored compounds has been intense local heating caused by partial absorption of the laser beam, but sample rotation has done much to alleviate these effects (cf. Section 2.2.4). It is also possible to defocus the laser beam.

7.1.4   Glasses and Quartz

One of the major techniques used in the study of glasses and glassy solids—along with IR, NMR, and EPR—has been Raman spectroscopy. A complete review of the application of spectroscopic approaches to borate, silicate, germanate, phosphate, arsenite, other oxides, chalcogenide glasses, and molten and glassy salts was published by Wong and Angell (30).

An example of the use of Raman spectroscopy to study glasses is shown in Fig. 7.2, from Exarhos and Risen (31). They used Raman spectroscopy to study the devitrification process. Alkali phosphate glasses were also examined by Fawcett et al. (32) and Rouse et al. (33). Raman and IR spectra for two crystalline phases of VPO_5 (A and B) were obtained, which greatly aided in the interpretation and band assignments of related VPO_5 glasses.

Structure of glasses in the systems calcium oxide-sodium oxide-diboron trioxide and magnesium oxide-sodium oxide-diboron trioxide were studied by Komjnendijk (34). Verweij (35) studied the actual reactions in a potassium carbonate-silica glass-forming batch and was able to identify the various glassy and crystalline phases that occurred during the reactions. Other structure studies have involved lithium titanium silicate and lead silicate glasses (36), binary chalcogenide glasses (37), and lithium oxide-iron (III) oxide-silica glasses (38).

An unusual application of Raman spectroscopy involved the detection of OH^- and H_2O impurities in the crystal lattice of α quartz. This work was important because defects such as these lower the mechanical quality of quartz. Walrafen and Luongo (39) conducted investigations of hydrothermal α quartz and obtained

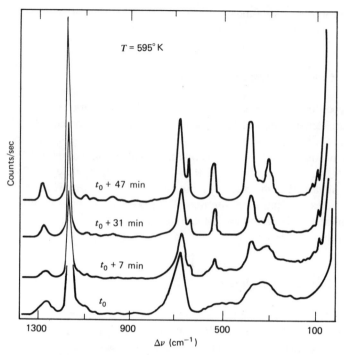

Figure 7.2 Raman spectra of $NaPO_3$ glass at 595°K showing the devitrification process (From ref. 31.)

valuable information involving the geometrical distribution as well as interactions between the OH^- and H_2O impurities.

Internal reflection spectroscopy (IRS) was combined with Raman spectroscopy to obtain a spectrum of α quartz. Raman-IRS spectra were first successfully recorded by Okeshoji et al. (40) for liquid carbon disulfide. Baptizmanskii et al. applied the method to α quartz, making it possible to compare the spectra of the surface layer and volume regions of the crystal (41). Thus the extension of the IRS method to Raman spectroscopy shows great promise in the study of the properties of thin surface layers of transparent crystals.

7.1.5 Carbon

The Raman spectrum of diamond was a great interest of C. V. Raman. The structural characterization of carbon has been studied extensively by Vidano and co-workers (42–44). The properties and behavior of carbon materials are strongly structure dependent. Raman spectroscopy can be used both to characterize virgin and treated-surface structures, and textures of various kinds of carbon materials, and to study their morphology and defects. An extensive survey of the Raman spectra of carbon materials has produced experimental evidence for at least five structure-sensitive bands (45). In addition to the 1580 cm^{-1} graphite line and the 1360 cm^{-1} disordered carbon line, which are always present, there is a disorder line at ~1620 cm^{-1} that is responsible for the apparent blue shift of the graphite line in very disordered carbon. There are also lines at 2700 and 2735 cm^{-1} that are strong in graphite and annealed carbon, but absent in disordered carbon. These additional lines enhance the capability of Raman spectroscopy to characterize carbon materials.

7.1.6 Sulfur Compounds

Pure sulfur and sulfur compounds have also been examined by Raman spectroscopy. Gautier and Debeau (46) studied single crystals of monoclinic β-S at various temperatures above and below the transition temperature. Janz et al. (47) investigated the S_3^{2-} anion from 125 to 580°C using BaS_3 as the model system. He and his co-workers also studied potassium and sodium polysulfides in the polycrystalline, molten, and glassy states (48, 49).

Solutions of alkali metals and sulfur in liquid ammonia and amines have been the subjects of numerous investigations by various techniques. The solutions are characterized by unusual colors and high conductivities, but the exact nature of the solvent or solute is not understood. The feasibility of using Raman spectroscopy to elucidate the nature of sulfur dissolved in amines was shown by Daly and Brown, who obtained the Raman spectra of rhombic sulfur dissolved in various primary amines (50) and secondary amines (51).

A very practical application of Raman spectroscopy involved the study of sulfur in wood bonding systems. The identification of CH_2—S bands indicated that the sulfur reacts with formaldehyde-urea copolymer present in the systems (52).

Another practical application involved the identification of sulfur deposits in bubbles in glass (53). Fig. 7.3 shows the Raman spectrum obtained from a

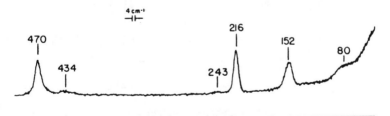

Figure 7.3   Raman scattering from deposit in glass bubble, 514.5 nm excitation. (From ref. 53.)

119

deposit on the surface of an inclusion from a clear soda-lime-silica glass plate. The lines clearly identify the deposit as elemental sulfur, and the lack of structure indicates it is in its polymeric form $S_\infty + S_8$ (54). Raman spectroscopy was also used to identify the gaseous contents of the same bubble and to observe a chemical reaction that changed the relative concentrations of the gases as the sample was heated.

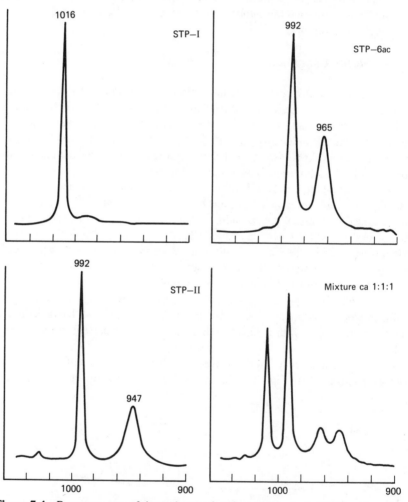

Figure 7.4   Raman spectra of three phases of sodium triphosphate (STP) and a mixture of all three phases. The data can be used for quantitative analysis. (From ref. 55.)

120

7.2 CHANGES OF STATE AND PHASE TRANSITIONS

Bus (55) applied Raman spectroscopy to quantitatively determine the amounts of three different phases of sodium triphosphate. As seen in Fig. 7.4, the different phases are easily distinguished, and quantitative analysis can be obtained with low standard deviations.

In the stable solid state, other spectroscopic studies have shown $PC\ell_5$ to have

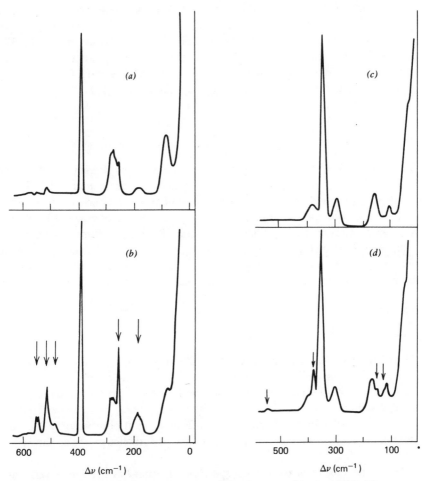

Figure 7.5 Gas phase Raman spectrum of $PC\ell_5$ at: *(a)* 160°C; *(b)* 220°C. The arrows indicate bands of $C\ell_2$ and $PC\ell_3$. Gas phase Raman spectra of $SbC\ell_5$ at: *(c)* 150°C; *(d)* 250°C. The arrows indicate bands of $C\ell_2$ and $SbC\ell_3$. (From ref. 56.)

121

the ionic structure $PC\ell_4^+ PC\ell_6^-$, but comparison of the spectra of the vapor and of a quenched metastable state showed they were both the simple molecular form $PC\ell_5$ (see Fig. 7.5) (56).

Cody and Darlington (9) described the very effective use of thermal analysis with Raman spectroscopy to follow interconversion of crystal structures in antimony oxides and lead carbonate. They also combined thermal analysis and Raman spectroscopy to provide a mechanistic understanding of the observed thermoanalytic phase transition behavior of Na_2SO_4 as a function of temperature.

The study of Raman spectra of oriented crystals as a function of temperature and pressure can yield much valuable information about the structural changes that accompany phase transitions. Ammonium chloride (57), sodium nitrate (58), and potassium selenate (59) are among the inorganic crystals that have been studied. Phase transitions in dicesium lithium iron hexacyanide, $Cs_2LiFe(CN)_6$, were also studied and compared to those of related salts (60).

7.3 MOLTEN SALTS

There have been extensive studies of molten salts by Raman spectroscopy (47–79), although this area has not been very active since the mid-1970s. Wilmhurst (61) discussed the need to be aware of complex ions and ion pairs when interpreting such spectra. A classic (pre-laser) study was carried out by Woodward et al. (62) on $GaC\ell_2$, showing from the Raman spectrum that the actual structure is $Ga[GaC\ell_4]$. Quist (63) has discussed cell designs for corrosive molten fluorides. He shows that a windowless cell, in which the liquid is held in place by its surface tension, can yield excellent Raman spectra.

7.4 WATER

There are many Raman spectroscopic studies of neat liquids, including several that focus on the effect of solutes on these liquids. (See (64) for a review.) However, water deserves a separate section in view of the thousands of papers on this subject alone. There are two major motivations: 1. to learn about the structure of water in all its many phases from spectral observations; 2. to understand aqueous solutions as viewed from the perturbation of the spectroscopic data by solutes.

While the subject of the Raman spectrum of water has been reviewed many times, the excellent review by Scherer (65) summarizes all pertinent data to 1977, including references to previous reviews.

An important approach used in the study of water is isotopic substitution. This allows three species to be readily measured, H_2O, D_2O, and HOD. HOD is particularly important as the difference in vibrational frequencies effectively uncouples the oscillators. In such spectroscopic studies one can then make independent OH and OD stretching measurements on the same system. The only major criticism of this approach is that some D_2O may be present even for dilute solutions of HOD in H_2O.

In examining spectra of HOD (66), polarization separation into isotropic and anisotropic components is useful. It helps to indicate the origin of shoulders and inflection points in the overall contour. The depolarization ratios measured from such data show that $\rho_{OH} > \rho_{OD}$. Scherer (66) has explained this, indeed, has shown that it is a valuable piece of data for looking at asymmetry in the weak oscillator coupling that persists in HOD. Comparable data have also been obtained for H_2O.

Scherer (66) has pointed out that the decomposition of the spectra into isotropic and anisotropic polarizabilities is important as well because of the previous contention (67) that the temperature-dependent Raman spectra (with polarizability unresolved) showed an isosbestic point. This was cited as evidence for two species. The isotropic and anisotropic spectra do not show this isosbestic point, leading to a conclusion that the observation is fortuitous.

Scherer has emphasized both the spectroscopic complexity of the water spectra and the difficulty in drawing conclusions from them. Such phenomena as anharmonicity, Fermi resonance, Evans holes (68), intermolecular coupling of vibrational modes, and the interaction of internal and external modes must be considered. Table 7.1, from Scherer (65), summarizes his conclusions regarding both the vibrational spectrum of water and the structural information that may be drawn from these data.

7.5 EQUILIBRIUM, DISSOCIATION, AND REDISTRIBUTION REACTIONS

The relatively weak Raman spectrum of water and the well-classified spectra of nonaqueous solvents for electrolytes have led to numerous investigators using Raman spectroscopy to study equilibria in electrolytic solutions. The entire subject up to 1976 has been reviewed by Irish and Brooker (69), and nearly 700 references in that review testify to the great quantity of work in this area.

Some of the studies of aqueous solutions are covered in other sections of the book. And applications are often covered in the biannual review in *Analytical Chemistry*.

The use of Raman spectroscopy to determine acid dissociation constants has been summarized (69). The results are shown in Table 7.2. While Raman spec-

Table 7.1   Conclusions Drawn from Vibrational Spectroscopic Studies of Water

Spectroscopic Conclusions

Fermi resonance between OH stretching and $2\nu_2$ states has been observed in the spectra of water in the gas phase, hydrogen bonding solvents, hydrates, ice, and the liquid state. The magnitude of the resonance interaction is small and causes Evans holes or weak shoulders to appear in the observed spectra, depending on the distance between, and widths of, the fundamental and overtone distributions.

Band-fitting procedures that approximate the Evans hole with two bands attributed to molecular species are invalid.

There is very little basis for fitting the uncoupled oscillator spectra of liquid water with any more than two bands. One band is associated with strongly bonded OH groups and the other with weakly bonded OH groups. The assumption of symmetric Gaussian band shapes for IR or Raman spectra has no theoretical or experimental basis.

Intermolecular coupling between adjacent water molecules is important in ice, amorphous ice, and liquid water. It affects the intensity distribution as a function of frequency in polarized Raman spectra and makes the uncoupled oscillator frequencies appear high relative to the observed IR and Raman band maxima. Since the strongly bonded OH of an asymmetrically hydrogen bonded water molecule is strongly coupled to its nearest neighbors, it is impossible to separate its distribution of frequencies from that due to the surrounding di-binded (tetra-coordinated) molecules. Consequently, it is probably erroneous to fit the H_2O water spectrum with any more than two bands, one with an Evans hole.

Normal-coordinate calculations that distinguish between symmetrically and asymmetrically hydrogen bonded water are helpful in interpreting the spectra of noncoupled H_2O in hydrogen bonding solvents, dimers, hydrates, and uncoupled HOD in all hydrogen bonded systems.

The change in dipole moment with OH bond stretch increases linearly with increasing hydrogen bond strength (decreasing uncoupled HOD oscillator frequency).

The ratio of longitudinal to transverse polarizability change with OH-bond stretch has a value of approximately 5.55 and is the same for liquid water at 0°C as it is for ice at -4°C. This ratio is higher for OH bonds that are weakly hydrogen bonded.

There are substantial anharmonicity differences between symmetrically and asymmetrically hydrogen bonded water molecules, differences that can influence the width of observed OH-stretching bands.

Isosbestic regions in the Raman spectra of water cannot be interpreted as evidence for different molecular species.

Disorder is responsible for the band width in liquid water and amorphous ice. The disorder probably arises from bent bonds, dipole-diple coupling, and interaction with translational and librational motions.

Structural Conclusions

Liquid water at low temperatures is similar to ice near the melting point or amorphous ice at $-180°C$. It is largely four-coordinated with a small amount of water that has one hydrogen atom that is only weakly hydrogen bonded.

The distortion of hydrogen bonding in the liquid at low temperatures is slightly greater than that in amorphous ice at $-180°C$, which is, again, slightly greater than in single crystal ice near the melting point. The halfwidths of the uncoupled oscillator bands are 250, 210, and 65 cm^{-1} for the OH stretch in these three systems, respectively.

There is evidence for nontetrahedral HOH valence angles in ice, with some orientational preference for c axis hydrogen atoms to form more linear hydrogen bonds.

Heating liquid water involves the formation of larger amounts of asymmetrically hydrogen bonded water molecules. At 400°C at a density of 1 g/cm^3, most of the water molecules are still weakly hydrogen bonded. At this point, broken hydrogen bonds are only produced by lowering the density.

The observation of broad bands in the spectra of single crystal ice near the melting point strongly suggests that the width of the corresponding bands in the spectra of liquid water is not due to the presence of a small number of distinctly differently hydrogen bonded species.

There is little need to invoke a two (or more) state model to interpret the IR and Raman spectra of water.

From ref. 65.

Table 7.2   Vibrational Spectral Studies of Acid Dissociation

Acid	Species Monitored	Raman (cm^{-1})	Ka
CH_2FCOOH	$CH_2FCO_2^-$	1340	$3.0 \pm 0.5 \times 10^{-3}$
CF_3COOH	$CF_3CO_2^-$	1438	4–8
$CC\ell_3COOH$	$CC\ell_3CO_2^-$	1348	2–5
		1344	3.2
CH_3SO_3H	$CH_3SO_3^-$	1046	83 ± 2
$C_2H_5SO_3H$	$C_2H_5SO_3^-$	1046	48 ± 2
$C_3H_7SO_3H$	$C_3H_7SO_3^-$	1046	34 ± 6
HSO_4^-	SO_4^{2-}	980	0.01
HNO_3	NO_3^-	1049	24

From ref. 69. See ref. 69 for sources of original data.

troscopy does not yield extremely high accuracy dissociation constants, it is useful in allowing the concentrations of a variety of species present in the solution to be measured over a wide range.

Raman spectroscopy has proven generally useful for studying dissociation and redistribution equilibria. This can be done in the gas phase as well. Fig. 7.5 shows the gas phase Raman spectra of $PC\ell_5$ and $SbC\ell_5$ at two different temperatures. As the temperature is raised, bands due to $PC\ell_3$, $SbC\ell_3$, and $C\ell_2$ appear in the spectra (56). Redistribution reactions of halogens, both in solution and in the gas phase, can also be studied conveniently by Raman spectroscopy.

7.6   COMPLEX IONS

Complex ions are another species that can be studied by Raman spectroscopy. These have been extensively discussed by Irish and Brooker (69). Solvent extraction techniques to isolate a particular complex species have been particularly useful and have been applied to systems including the chloro- and bromo-complexes of mercury (II) (70), cadmium (II) (71), thallium (72), and arsenic (73). Tytko and Schonfeld (74) investigated the relation between solid isopolymolybdates and their ions in solution and found that the octamolybdate ion $Mo_8O_{26}^{4-}$ is not present in appreciable quantities in solution at room temperature, contrary to previously published results.

Aqueous solutions of NaH_2PO_4 were investigated by Steger et al. (75). A mathematical separation of IR bands between 700 and 1500 cm^{-1} and a study of the Raman spectra indicated the anions associated into chains that had a line group symmetry isomorphous with C_{2h}.

7.7 RESONANCE RAMAN SPECTRA OF INORGANIC MOLECULES

An area that is attracting an increasing amount of attention is the resonance Raman spectra of inorganic molecules. Comprehensive reviews have been written by Clark and Stewart (76, 11). As indicated previously, the resonance Raman effect (RRE) arises when $(\nu_e - \nu_0) \approx 0$ (ν_0 = exciting frequency, ν_e = the frequency of the lowest allowed electronic transition of the molecule), and it is characterized by high intensity overtone progressions of a totally symmetric fundamental. This is shown for the $Mo_2C\ell_8^{4-}$ ion in Fig. 7.6 (77). The study of this ion is particularly interesting because it is known that the transition at about 19,000 cm^{-1} involves excitation of the δ-electron of the so-called quadruple Mo—Mo bond (78, 79). By bringing the exciting frequency into coincidence with this band, the totally symmetric vibration ν(Mo—Mo) is resonance enhanced, confirming the close relationship between absorption spectroscopy and RRE.

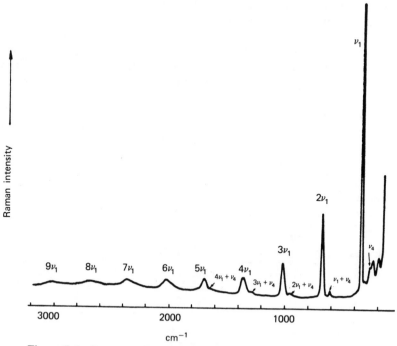

Figure 7.6 Resonance Raman spectrum of $Cs_4Mo_2C\ell_8$. (From ref. 77.)

Clark and Mitchell studied the resonance and preresonance spectra of titanium tetraiodide (80) and were able to determine the harmonic frequencies and anharmonicity constants for the $v_1(a_1)$ fundamental.

An unusual application of RRE is the identification of the sulfur species present in the deep blue mineral ultramarine, which is essentially a sodium aluminosilicate with the idealized formula $(Na_8A\ell_6Si_6O_{24}S_4)_n$ (81). Resonance Raman has also been used as a probe of the annealing behavior of Cu_2O implanted with 180 keV Cd^{2+} ions (82). As annealing progressed, large changes were observed in the resonance Raman spectra. When the annealing stage at ~250°C was reached, >99% of the damage caused by implantations was erased. Other interesting applications can be found in Clark's review article (76) on resonance Raman spectra of inorganic ions and molecules.

7.8 MINERALS

Raman spectroscopy can be used very effectively for diagnostic purposes on minerals. In general, the Raman spectra of minerals are simpler in appearance than the comparable IR spectra, the bands are sharper, and the interpretation simpler. The position of v_1, the totally symmetric (MO_n) stretch, and normally the strongest and sharpest band in the spectrum, is a useful preliminary guide to mineral identification. Griffith (6, 83) recorded the Raman spectra of a number of carbonate, phosphate, arsenate, vanadate, niobate, tantalate, sulfate, chromate, molybdate, and tungstate minerals. It appears from a study of these spectra that for minerals containing MO_3 and MO_4 groups, a "fingerprint" approach is possible much the same as for IR mineral spectra. Table 7.3 lists the positions of the strongest Raman bands (i.e., v_1, totally symmetric stretch) in various MO_3

Table 7.3 Positions of Strongest Raman Bands (i.e., v_1, Totally Symmetric Stretch) in Various MO_3 and MO_4 Minerals

Mineral	Raman Band (cm^{-1})	Mineral	Raman Band (cm^{-1})
Carbonates	1100	Arsenates	830
Nitrates	1050	Chromates	880
Silicates	880–1000	Molybdates	870
Sulphates	990	Tungstates	910
Phosphates	960	Vanadates	830

From ref. 6.

and MO_4 minerals. But comparatively little is known about the Raman spectra of minerals thus far, leaving a vast field of potentially exciting investigations.

REFERENCES

1. M. M. Sushchinskii, *Raman Spectra of Molecules and Crystals,* Keter, New York, 1972.
2. K. Nakamoto, *Infrared and Raman Spectra of Inorganic and Coordination Compounds,* 3rd ed., Wiley, New York, 1978.
3. D. M. Adams, *Metal-Ligand and Related Vibrations,* St. Martin's Press, New York, 1968.
4. R. J. H. Clark, *The Spex Speaker* **18,** 1 (1973).
5. R. L. Carter, in *Infrared and Raman Spectroscopy,* Vol. 1, Part A, E. G. Brame and J. G. Grasselli, Eds., Dekker, New York, 1977.
6. W. P. Griffith, in *The Infra-Red Spectra of Minerals,* V. C. Farmer, Ed., Mineralogical Society, London, 1974.
7. G. Turrell, *Infrared and Raman Spectra of Crystals,* Academic, New York, 1972.
8. W. G. Fateley, F. R. Dollish, N. T. McDevitt, and F. F. Bentley, *Infrared and Raman Selection Rules for Molecular and Lattice Vibrations; The Correlation Method,* Wiley, New York, 1972.
9. C. A. Cody and R. K. Darlington, *The Spex Speaker* **25,** 1 (1980).
10. P. P. Shorygin, *Russ. Chem. Rev.* **40,** 367 (1974).
11. R. J. H. Clark and B. Stewart, in *Structure and Bonding,* Vol. 36, J. D. Dunitz et al., Eds. Springer-Verlag, New York, 1979.
12. E. J. Baran, *Z. Anorg. Allg. Chem.* **442,** 112 (1978).
13. E. J. Baran, *Monatsch. Chem.* **106,** 1 (1975).
14. Y. S. Jain, V. K. Kapoor, and H. D. Bist, *Appl. Spectrosc.* **30,** 440 (1976).
15. R. L. Carter and C. E. Bricker, *Spectrochim. Acta, Part A* **30,** 1793 (1974).
16. R. Carter and L. O'Hare, *Appl. Spectrosc.* **30,** 187 (1976).
17. J. San Filippo, Jr. and H. J. Sniadoch, *Inorg. Chem.* **12,** 2326 (1973).
18. J. G. Grasselli, M. A. Hazle, and L. E. Wolfram, in *Molecular Spectroscopy,* Q. West, Ed., Heyden, New York, 1977.
19. C. Angell, paper presented at Eastern Analytical Symposium, Atlantic City, N.J., November 1972.
20. M. A. Py, P. E. Schmid, and J. T. Vallin, *Nuovo Cimento Soc. Ital. Fis. B.* **38,** 271 (1977).
21. P. J. Miller, *Spectrochim. Acta, Part A* **27,** 957 (1971).
22. H. Knözinger and H. Jeziorowski, *J. Phys. Chem.* **82,** 2002 (1978).
23. M. Liegeois-Duyckaerts, *Spectrochim. Acta, Part A* **31,** 1585 (1975).

24. M. Liegeois-Duyckaerts and P. Tarte, *Spectrochim. Acta, Part A* **30,** 1771 (1974).

25. M. T. Paques-Ledent, *Spectrochim. Acta, Part A* **32,** 1339 (1976).

26. C. N. R. Rao, *Indian J. Pure Appl. Phys.* **16,** 277 (1978).

27. G. R. Wilkinson, *Mol. Spectrosc.—Chem. Rev.* **3,** 433 (1975).

28. J. R. Allkins and P. J. Hendra, *J. Chem. Soc. A,* 1235 (1967).

29. R. J. H. Clark, G. Natile, U. Belluco, L. Cattalini, and C. Filipin, *J. Chem. Soc. A,* 659 (1970).

30. J. Wong and C. A. Angell, *Appl. Spectrosc. Rev.* **4,** 155 (1971).

31. G. J. Exarhos and W. M. Risen, Jr., *J. Amer. Ceram. Soc.* **57,** 401 (1974).

32. V. Fawcett, D. A. Long, and L. Taylor, *Proc. 5th Int. Conf. Raman Spectrosc.,* 1976, Heyden, London, p. 112.

33. G. B. Rouse, P. J. Miller, and W. M. Risen, Jr., *J. Non-Cryst. Solids* **28,** 193 (1978).

34. W. L. Komjnendijk, *Phys. Chem. Glasses* **17,** 205 (1976).

35. H. Verweij, H. Van den Boom, and R. Breemer, *J. Amer. Ceram. Soc.* **61,** 118 (1978).

36. T. Furukawa, *Diss. Abstr. Int. B* **38:** 12, Pt 1, 6091 (1978).

37. G. Lucovsky, F. L. Galeener, R. H. Geils, and R. C. Keezer, *Struct. Non-Cryst. Mater., Proc. Symp. 1976,* published 1977, p. 127.

38. S. A. Brawer and W. B. White, *J. Mater. Sci.* **13,** 1907 (1978).

39. G. E. Walrafen and J. P. Luongo, *The Spex Speaker* **20,** 1 (1975).

40. T. Okeshoji, Y. Ono, and T. Mizuno, *Appl. Opt.* **12,** 2236 (1973).

41. V. V. Baptizmanskii, I. I. Novak, and A. E. Chmel, *Opt. Spectrosc. (USSR)* **43,** 106 (1977).

42. R. Vidano and D. B. Fischbach, "Raman Spectroscopy of Carbon Materials: I. New Spectral Lines. II. Characterization of Materials," paper presented at American Ceramic Society Pacific Coast Regional Meeting, San Francisco, Calif., Oct. 31–Nov. 3, 1976.

43. T. G. Miller, D. B. Fischbach, and J. M. Macklin, *Ext. Abstr. Program-Bienn. Conf. Carbon* **12,** 105 (1975).

44. R. Vidano and D. B. Fischbach, *Ext. Abstr. Program-Bienn. Conf. Carbon* **13,** 272 (1977).

45. R. Vidano and D. B. Fischbach, *J. Amer. Ceram. Soc.* **61,** 13 (1978).

46. G. Gautier and M. Debeau, *Spectrochim. Acta, Part A* **32,** 10007 (1976).

47. G. J. Janz, E. Roduner, J. W. Coutts, and J. R. Downey, Jr., *Inorg. Chem.* **15,** 1951 (1976).

48. G. J. Janz, J. W. Coutts, J. R. Downey, Jr., and E. Roduner, *Inorg. Chem.* **15,** 1755 (1976).

49. G. J. Janz, J. R. Downey, Jr., E. Roduner, G. J. Wasilczyk, J. W. Coutts, and A. Eluard, *Inorg. Chem* **15**, 1759 (1976).

50. F. P. Daly and C. W. Brown, *J. Phys. Chem.* **77**, 1859 (1973).

51. F. P. Daly and C. W. Brown, *J. Phys. Chem.* **80**, 480 (1976).

52. B. Meyer and W. E. Johns, *Holzforschung* **32**, 102 (1978).

53. G. J. Rosasco and J. H. Simmons, *Amer. Ceram. Soc. Bull.* **54**, 590 (1975).

54. A. T. Ward, *J. Phys. Chem.* **72**, 4133 (1968).

55. J. Bus, *Anal. Chem.* **46**, 1824 (1974).

56. G. A. Ozin, *Prog. Inorg. Chem.* **14**, 173 (1971); I. R. Beattie, *Chem. Br.* **4**:3, 347 (1967).

57. D. A. Long, *16th Colloq. Spectrosc. Int., Plenary Lect. Rep., 1971*, Adam Hilger, Ltd., London, published 1972.

58. A. D. Prasad Rao, R. S. Katiyar and S. P. S. Porto, in *Advances in Raman Spectroscopy*, J. P. Mathieu, Ed., Heyden, London, 1973.

59. C. Caville, V. Fawcett, and D. A. Long, *Proc. 5th Int. Conf. Raman Spectrosc.*, 1976, Heyden, London, p. 626.

60. B. I. Swanson, B. C. Lucas, and R. R. Ryan, *J. Chem. Phys.* **69**, 4328 (1978).

61. J. K. Wilmhurst, *J. Chem. Phys.* **39**, 1779 (1963).

62. L. A. Woodward, G. Garton, and H. L. Roberts, *J. Chem. Soc.*, 3723 (1956).

63. A. S. Quist, *Appl. Spectrosc.* **25**, 80 (1971).

64. D. E. Irish, in *Physical Chemistry of Organic Solvent Systems*, A. K. Covington and T. Dickenson, Eds., Plenum, New York, 1973, p. 433.

65. J. R. Scherer, in *Advances in Infrared and Raman Spectroscopy*, Vol. 5, R. E. Hester and J. H. Clarke, Eds., Heyden, London, 1978, p. 149.

66. J. R. Scherer, M. K. Go, and S. Kint, *J. Phys. Chem.* **78**, 1304 (1974).

67. G. E. Walrafen in *Water*, Vol. 1, F. Franks, Ed., Plenum, New York, 1972, chap. 5.

68. J. R. Scherer, M. K. Go, and S. Kint, *J. Phys. Chem.* **77**, 2108 (1973).

69. D. E. Irish and M. H. Brooker, in *Advances in Infrared and Raman Spectroscopy*, Vol. 2, R. J. H. Clark and R. E. Hester, Eds., Heyden, London, pp. 212–311.

70. J. E. D. Davies and D. A. Long, *J. Chem. Soc. A*, 2564, (1968).

71. J. E. D. Davies and D. A. Long, *J. Chem. Soc. A*, 2054 (1968).

72. J. E. D. Davies and D. A. Long, *J. Chem. Soc. A*, 2050 (1968).

73. J. E. D. Davies and D. A. Long, *J. Chem. Soc. A*, 1761 (1968).

74. K. H. Tytko and B. Schonfeld, *Z. Naturforsch, Teil B* **30**, 471 (1975).

75. V. E. Steger, K. Herzog, and J. Klosowski, *Z. Anorg. Allg. Chem.* **432**, 42 (1977).

76. R. J. H. Clark, in *Advances in Infrared and Raman Spectroscopy*, Vol. 1, R. J. H. Clark and R. E. Hester, Eds., Heyden, London, 1975.

77. R. J. H. Clark and M. L. Franks, *J. Chem. Soc., Chem. Commun.* **9,** 316 (1974).

78. F. A. Cotton and C. B. Harris, *Inorg. Chem.* **6,** 924 (1967).

79. C. D. Cowman and H. B. Gray, *J. Amer. Chem. Soc.* **95,** 8177 (1973).

80. R. J. H. Clark and P. D. Mitchell, *J. Amer. Chem. Soc.* **95,** 8300 (1973).

81. R. J. H. Clark and M. L. Franks, *Chem. Phys. Lett.* **34,** 69 (1975).

82. J. F. Hesse, S. C. Abbi, and A. Compaan, *J. Appl. Phys.* **47,** 5467 (1976).

83. W. P. Griffith, *J. Chem. Soc. A,* 286 (1970).

Chapter Eight

Surfaces and Catalysts

The study of surfaces and surface phenomena is very important to the industrial chemical world in the fields of coatings, metals, corrosion chemistry, colloid chemistry, and catalysis. Chemistry at surfaces is different from chemistry of bulk materials, and analytical methods such as ESCA (electron spectroscopy for chemical analysis), Auger spectroscopy, electron microscopy, and LEED (low energy electron diffraction) are all useful in characterizing various aspects of surface structure. Raman and IR spectroscopy are included in this discussion because vibrational spectra reflect so well the chemical and physical changes in a molecule as affected by its environment. Thus they are extremely valuable for studying adsorbed species to obtain information about the type and nature of the active sites of surfaces or to elucidate the structure and bonding of the adsorbates. They also provide structural information on bulk catalysts.

Infrared spectroscopy often achieves a high signal-to-noise ratio and has been extensively used in studies of adsorbed species (1), but the intense bands of common oxide substrates often interfere with much of the interesting spectral region. Raman spectroscopy offers a decided advantage in such cases, particularly for molecules that are good scatterers. In addition, its water transparency allows the study of solid-aqueous interfaces. Many publications, especially since 1967, have described the application of Raman techniques to study surface-adsorbent interactions, the chemical nature of surface films and layers, and even the study of electrodes beneath electrolytes in electrochemical cells.

Raman scattering is usually quite weak on a per molecule basis, and one would thus be surprised to find that Raman spectroscopy could be used to study low surface coverages, such as monolayers. In the first section of this chapter we describe the results from conventionally weak Raman spectra on surfaces, including attempts to enhance these by resonance Raman scattering. We then discuss the rather startling observation of very intense Raman scattering from certain molecular species adsorbed on metals. In such cases one can observe enhancement of Raman intensity of 10^6.

8.1 CONVENTIONAL RAMAN SPECTRA OF SURFACES

There are several reviews available in the literature of solid-solid, solid-gas interfaces, among which the most useful are ones by Paul and Hendra (2) and Cooney et al. (3). Good examples of solid-solid interfaces were given by Buechler and Turkevich (4), who interpreted the Raman spectra of molybdenum trioxide on porous Vycor, uranium oxide on porous glass, and platinum on silica.

Solid-gas interactions have been studied by Stencel and Bradley, who exposed the surface of Ni (III) to CO, H_2, and O_2 in the 300°C region (5). In the Raman spectrum the changes of a major band at 80 cm^{-1} upon adsorption of the gases were shown and structures of the possible adsorbed species were discussed. This work was extended (6) by the development of an ultrahigh vacuum chamber, which enabled sample temperatures from $-85°$ to 600°C to be obtained, along with surface-cleaning capabilities. Another useful device for gas-solid reaction mechanism studies was developed by Kagel et al. (7), who constructed a high pressure cell capable of withstanding pressures in excess of 650 psig and evacuable to 5 torr. The entire cell is enclosed in a small furnace for elevated temperature studies.

There are two main classifications of substrates which are of catalytic importance. The first of these constitutes oxides of aluminum and silicon, and the important adsorbents derived from them—the silica-alumina cracking catalysts and the zeolites, which are crystalline silica-aluminas with replaceable metal cations. Morrow (8) obtained the Raman spectrum of chemisorbed methanol on silica and compared it to the IR spectrum. Yamamoto and Yamada (9) examined the spectral changes of the three diazines: pyrimidine, pyrazine, and pyridazine absorbed on SiO_2. A series of articles were published by Tam et al. (10–13) on the vibrational spectra of acetylene on A-type zeolites and pyrazine, acetylene, dimethylacetylene, and cyclopropane on X-type zeolites. The dependency of the Raman line shift on the nature of the zeolitic cation is discussed in each case.

Similar studies of adsorbed oxygen, nitrogen, benzene, and cyclopropane (14–16) used shifts in the Raman spectra of the adsorbed molecules to probe the electrostatic fields within the zeolite super cage.

Chemisorption and reactions on zeolite molecular sieves were studied by Trotter (17). The advantage of observing low frequency bands in the Raman spectrum in the presence of the zeolite framework was critical because it allowed a kinetic comparison between silver complex formation with benzotriazole at 793 cm^{-1} and the surface formation of N—Ag—N bonds at 136 cm^{-1}.

134

In addition to the silica and alumina adsorbents, the second major type of support consists of fine metal particles (18, 19) usually stabilized by decomposition on supports of the high surface area oxide types just described. Krasser et al. (20) observed the Raman scattering of hydrogen chemisorbed on silica-supported nickel.

The most extensive studies of surface interactions have involved the adsorption of pyridine on various adsorbents (21–24). One of the reasons for the choice of pyridine is because its vibrational modes shift in very well-defined ways dependent upon the site to which it is attached. An excellent review and list of references can be found in the Paul and Hendra article (2). Table 8.1 shows how sensitive these Raman bands of pyridine are to its different bonding environments since the pyridine can chemisorb on the surface by donation of the lone pair of electrons on the nitrogen atom to Lewis acid sites (a coordinate bond); it can hydrogen bond to surface OH groups; or it can form the pyridinium cation (Brönsted acid sites) by complete proton abstraction from surface OH groups.

In recent years there has been a very high interest in the nature of reactions that occur at electrode-electrolyte surfaces (25). An understanding of these reaction mechanisms is extremely important during the current energy crisis because of the quest for new kinds of fuel cells and storage batteries. Because of the capability of obtaining Raman spectra of aqueous solutions, Raman has been the major technique used to monitor changes at mercury (26), silver (21), copper (27, 28), and platinum (29) electrodes as a function of the applied potential.

Fleischmann et al. (28) selected pyridine as the molecule to probe the nature of the adsorption sites on the electrode surfaces. Fig. 8.1 shows the Raman spectra of pyridine in solution and at a silver electrode at various potentials in 0.05 pyridine/0.1 M aqueous potassium chloride. The bands at ~ 1008 and ~ 1036 cm^{-1} pass through an intensity maximum close to the zero of potential change. With increasing negative potential, there is a significant change in the frequency of the ring breathing mode and a new band is observed at 1025

Table 8.1 Variations in the Frequencies (cm^{-1}) of the C-H Stretch and Ring-Breathing Vibrations of Pyridine with the Nature of the Bonding

System	Bonding	Ring-Breathing Vibrations		C-H stretch
Pyridine		991	1031	3060
10% pyridine in CH$_3$OH	H-bond	996	1030	3062
10% pyridine in H$_2$O	H-bond	1004	1036	3074
10% pyridine in HCl	Brönsted	1010	1029	3109
Zn (pyridine)$_2$Cl$_2$	Lewis	1023	1047	3075, 3085

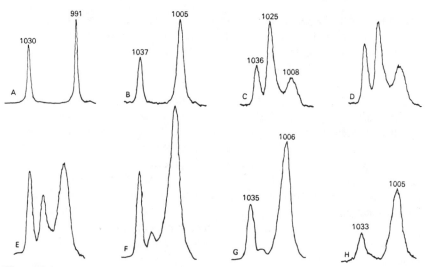

Figure 8.1 Raman spectra of pyridine in solution and at the silver electrode: *(a)* liquid pyridine; *(b)* 0.05 M aqueous pyridine; *(c)* silver electrode in 0.1 M potassium chloride + 0.05 M pyridine solution at 0 V (SCE); *(d)* −0.2 V; *(e)* −0.4 V; *(f)* −0.6 V; *(g)* −0.8 V; *(h)* −1.0 V. (From ref. 2.)

cm^{-1}. This latter band had previously been interpreted (21) as being due to pyridine directly coordinated to silver, the other bands being due to pyridine adsorbed through a polarized layer of water molecules. However, some of the changes could also be due to the interaction of pyridine with adsorbed carboxylate groups on the surface of the electrode. Recently, Cheng and Schrader (30, 31) have used pyridine adsorption as a probe of the surface acidity and structure of supported molybdate and cobalt molybdate catalysts. A change in the surface structure with molybdenumloading and with the addition of cobalt were noted. A new form of pyridine adsorption, characterized by a band at 1008 cm^{-1} in the Raman spectrum, was observed and assigned to the adsorption on aggregated molybdenum species at intermediate levels of molybdenum loading.

The development of resonance Raman techniques has proved invaluable in the study of surfaces (32–34). Yamamoto and Yamada (35) discussed the structures and formation mechanisms of carbonium ions on porous Vycor glass. Takenaka and Nakanaga (36) devised a new method of total reflection of the exciting line to record the resonance Raman spectra of monolayers adsorbed at the interface between carbon tetrachloride and an aqueous solution of a surfactant and a dye. Jeanmaire and co-workers (37) successfully interfaced electrochemical radical ion generation with resonance Raman spectroscopy to obtain spectra for the

136

electrogenerated monoanion radical of tetracyanoquinodimethane upon excitation of its lowest $^2B_{1u}$ excited state (37–40). Thus resonance Raman spectroscopy has proven its utility as a molecularly specific monitor for electrochemical processes and an important tool for the characterization of charge-transfer solids.

Raman surface spectroscopy is important to the industrial world, for example, in the study of thin films and coatings (41, 42), insoluble monolayers spread on a water surface (43), or iron oxide films formed on metal surfaces (to facilitate the use of Raman spectroscopy for *in situ* analysis of corrosion) (44). It has also been used in the biological field, for example, to study molecular orientation at interfaces of polypeptide monolayers (45). So, although Raman is a relatively new technique in surface science, it has already proven itself of immense value, especially in those areas of "real world" surfaces—dirty and heterogeneous, containing mixtures of corrosion products, grease, oxides, and so on. This is in contrast to ESCA, LEED, and Auger spectroscopy, which need chemically clean surfaces, high vacuum conditions and samples treated in specially prescribed ways.

8.2   SURFACE ENHANCED RAMAN SCATTERING

Surface enhanced Raman scattering (SERS) must be viewed as an independent technique for the study of surface adsorption. The original observation of the phenomenon was an intensity enhancement of 10^6 in the Raman spectrum of pyridine on a silver electrode, after an electrochemical oxidation reduction cycle had taken place (21, 46–49).

Subsequent to initial observations, intensity enhancements have been observed for a wide variety of adsorbed molecules, (eg., CN^-, $CO_3^=$, CO, benzoic acid, isonicotinic acid, crystal violet) on several metals (Cu, Ag, Au) (50–52).

There have been many explorations of other experimental aspects of this phenomenon. Apparently scattering does not follow a v^4 dependence (49, 53, 54) and may show a maximum intensity enhancement at certain frequencies. There may be a broad continuum in addition to the enhanced Raman scattering (50, 55). Surface roughness seems to be important to the enhancement, and may be the role of the electrochemical cycling (47, 56–59). This work has been reviewed in detail by Burstein and Chen (60).

There have also been several attempts to develop theoretical models that predict the enhancement. Early ideas centered on a resonance Raman effect of some sort. One possibility raised was that this involved interaction with a metal absorption, the surface plasmon. Another model is the coupling of a molecular dipole and its image in the metal, leading to a greater polarizability. Still other attempts have tried to quantitate the enhancement to be expected from surface

137

roughness (34, 55–57, 60–63, 66, 67). Recent reports on catalytically interesting molecules, such as cyanopyridines (64) and ethylene and propylene (65), offer intriguing evidence in support of various SERS models. While the image dipole model seems to make many correct predictions, the overall theoretical approach to successfully explain these phenomena is not yet clear.

8.3   CATALYST RESEARCH

Raman spectroscopy has had an important role in industrial catalyst research, particularly in the study of selective oxidation of hydrocarbons by transition metal oxide catalysts. An impressive body of literature has developed over the last few years exploring the mechanisms of the reactions and delving into just how a heterogeneous catalyst works (68, 69). Most modern spectroscopic tools have been used to characterize the surface and bulk structures of these complex oxide structures. Raman spectroscopy has emerged as one of the most useful methods.

Molybdate catalysts have been used for many chemical reactions, including the selective oxidation of methanol to formaldehyde and the selective ammoxidation of propylene to acrylonitrile. In all cases, it has been recognized that the surface structure of the catalyst plays a vital role in the selectivity and activity of the catalyst (70–73). The nature of the catalyst surface and the reaction mechanisms can be probed with studies of adsorbed species, using both IR and Raman spectroscopy. However, it is the catalyst structure itself that governs the reaction; therefore, the examples in this section are concerned with the structural changes that occur.

8.4   TECHNIQUES FOR STUDYING CATALYSTS

Most of the transition metal oxides are colored solids ranging from light yellow or green to brown or black. Therefore, in order to obtain good Raman spectra, it is necessary to carefully consider and control some of the instrumental parameters. Most investigators utilize a rotating cell in order to prevent decomposition or phase changes in the samples. It is also possible to obtain the Raman spectra of industrial catalysts and mixed transition metal oxides without a rotating cell by carefully controlling the microscope objective setting and laser power at the sample (74).

An interesting publication by Payen et al. (75) reports a Raman spectroscopic study of some cobalt molybdenum catalysts using the laser microprobe. As discussed in Section 2.2.3, this instrument has a spatial resolution of the laser beam of approximately 1 μm. Only 3 mW of laser power at the sample was used. The authors observed segregation of phases on the alumina supports, in agreement with ESCA data. The special design of this instrument's collection optics permits the use of such low power while still providing good spectral data.

In hydrodesulfurization, commercial catalysts are supported on alumina, and several investigators have reported the interaction between the support and the cobalt molybdenum catalyst (76–79). But very often the literature is not consistent on spectral data or structures of model compound metal oxides. In Fig. 8.2, the spectra of ferric molybdate as obtained by Villa and co-workers (72) and Medema and co-workers (76) are compared. They differ markedly, primarily as a result of the method of preparation. Grasselli et al. report data (74) that agrees with that of Villa et al.

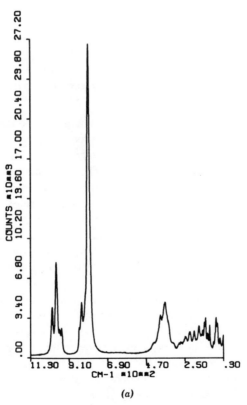

(a)

Figure 8.2 Raman spectra of $Fe_2(MoO_4)_3$. (a) Sohio results [agree with Villa et al. (72)]; (b) Medema et al. results (76).

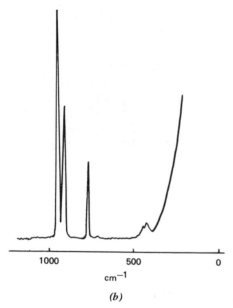

(b) **Figure 8.2** *(continued)*

As previously mentioned, perhaps one of the most important aids in obtaining good Raman spectra is the use of a data processing system with the Raman spectrometer. The capability to signal average in order to optimize signal-to-noise ratio, to mathematically minimize background interference, or to subtract spectra at will, along with accurate frequency and intensity output, make the computer an indispensable tool for Raman spectroscopy of catalysts. Fig. 8.3 shows a striking example of this capability reported by Grasselli et al. (74). It is the spectrum of molybdenum dioxide, MoO_2, a black solid. Fig. 8.3*a* shows a single scan, Fig. 8.3*b* shows 150 scans, and Fig. 8.3*c* shows the final spectrum, 150 scans smoothed and baseline corrected. The spectrum was obtained without sample rotation.

In previous work Grasselli et al. (80) described the application of a heated cell to follow phase transformations in molybdate catalysts. Since most industrial catalysts operate at elevated temperatures, it is very meaningful to examine the structures of the solids at these working temperatures. For example, bismuth molybdate is known to exist in three modifications, the α, β, and γ phases. The γ phase of Bi_2MoO_6 is a metastable compound with an x-ray diagram similar to that of the mineral koechlinite (81). Heating this compound to temperatures in excess of 660°C in air produces an irreversible transition to the γ' phase (69). Fig. 8.4 shows the transition from γ to γ' for bismuth molybdate as followed

140

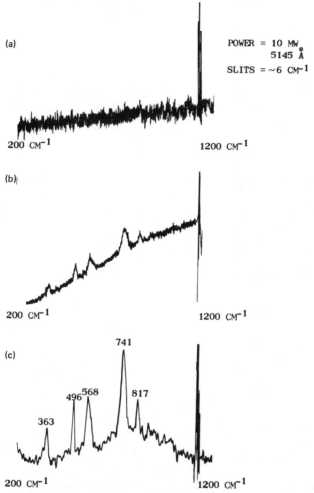

Figure 8.3 Raman spectra of MoO_2, a black powder: *(a)* single scan; *(b)* 150 scans, raw data; *(c)* 150 scans, smoothed and baseline corrected.

in the heated cell. From these data it appears that the onset of the γ' modification actually begins around 600°C, rather than the 650°C temperature reported in the literature. Such observations may be important when considering the relative activity of the various phases in catalysis, for example, in the oxidative dehydrogenation of 1-butene to butadiene where the γ phase is considered the active one.

141

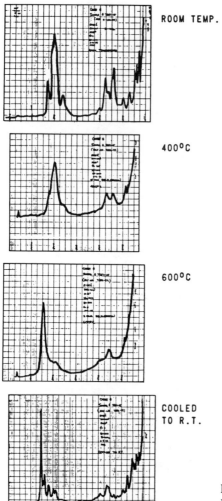

ROOM TEMP.

400°C

600°C

COOLED
TO R.T.

Figure 8.4 Phase transition, γ to γ', for Bi_2MoO_6.

8.5 SURFACE VERSUS BULK STRUCTURES

The importance of the surface structure in oxide catalysts has already been mentioned, not only in terms of identification of active sites for hydrocarbon adsorption in mechanism studies, but also because the structure of the surface in relation to the bulk influences the redox capability of the catalyst. Boudeville and co-workers (82) have recently emphasized the importance of the surface

142

structure relative to the selectivity in olefin oxidation on antimony-tin-oxygen catalysts. Correlations between x-ray photoelectron spectroscopy data and the catalytic properties of these materials were reported. Segregated phases that exist at the surface of the catalyst were detected.

A backscattering technique was used to provide direct information on the composition and structure of surface oxides formed on alloys (83). Spectra from inner regions also were measured using grazing-angle techniques. It was found that the presence of impurities and minor alloying constituents in the substrates have a marked effect on the results.

Raman spectra of alumina-supported rhenium oxide metathesis catalyst indicated that rhenium is present as tetrahedral ReO_4^- on the catalyst surfaces (84). Results suggested that these surface species are dynamically distorted by the carrier or surface hydroxyl groups.

The reduction of ferric molybdate was studied in a heated Raman cell to gain some insight into surface versus bulk structures. A parallel study was conducted using ESCA (74). The strongest band in the Raman spectrum of $Fe_2(MoO_4)_3$ is at 780 cm^{-1}, due to a Mo—O—Mo linkage (Fig.8.2). Fig. 8.5 shows successive scans of the 780 cm^{-1} band monitored at 3 min intervals. By examining these scans, it is clear that ferric molybdate initially undergoes a rapid reduction. This is followed by a slow, possibly first-order reduction, which is a function of the temperature. The sample reduces to ferrous molybdate at the end of the reduction. The ESCA data showed that only the molybdenum on the surface of the oxide is reduced. Thus, the combination of Raman and ESCA is invaluable for relating surface to bulk structure changes in catalysts.

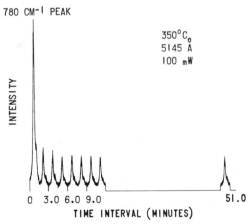

Figure 8.5   Reduction of $Fe_2(MoO_4)_3$ with propylene.

REFERENCES

1. N. Sheppard and T. T. Ngurjen, in *Advances in Infrared and Raman Spectroscopy,* Vol. 5, R. J. H. Clark and R. Hester, Eds., Heyden, London, 1978, p. 67.

2. R. L. Paul and P. J. Hendra, *Miner. Sci. Eng.* **8,** 171 (1976).

3. R. P. Cooney, G. Curthoys, and N. T. Tam, *Adv. Catal.* **24,** 293 (1975).

4. E. Buechler and J. Turkevich, *J. Phys. Chem.* **76,** 2325 (1972).

5. J. M. Stencel and E. B. Bradley, *Spectrosc. Lett.* **11,** 563 (1978).

6. J. M. Stencel, D. M. Noland, E. B. Bradley, and C. A. Frenzel, *Rev. Sci. Instrum.* **49,** 1163 (1978).

7. R. O. Kagel, R. A. Koster, and W. T. Allen, *Appl. Spectrosc.* **30,** 350 (1976).

8. B. A. Morrow, *J. Phys. Chem.* **81,** 2663 (1977).

9. Y. Yamamoto and H. Yamada, *Bull. Chem. Soc. Jap.* **51,** 3063 (1978).

10. N. Tam, R. P. Cooney, and G. Curthoys, *J. Chem. Soc., Faraday Trans. 1* **72,** 2591 (1976).

11. N. Tam, R. P. Cooney, and G. Curthoys, *J. Chem Soc., Faraday Trans. 1* **72,** 2592 (1976).

12. N. Tam and R. P. Cooney, *J. Chem. Soc., Faraday Trans. 1* **72,** 2598 (1976).

13. N. Tam, P. Tsai, and R. P. Cooney, *Aust. J. Chem.* **31,** 255 (1978).

14. D. D. Saperstein and A. J. Rein, *J. Phys. Chem.* **81,** 2134 (1977).

15. J. J. Freeman and M. L. Unland, *J. Catal.* **54,** 183 (1978).

16. T. T. Nguyen, P. Tsai, and R. P. Cooney, *Aust. J. Chem.* **31,** 255 (1978).

17. P. J. Trotter, *J. Phys. Chem.* **82,** 2396 (1978).

18. J. Heaviside, P. J. Hendra, P. Tsai, and R. P. Cooney, *J. Chem. Soc., Faraday Trans. 1* **74,** 2542 (1978).

19. W. Krasser, A. Fadini, and A. Renouprez, *J. Mol. Struct.* **60,** 427 (1980).

20. W. Krasser and A. J. Renouprez, *J. Raman Spectrosc.* **8,** 92 (1979).

21. M. Fleishmann, P. J. Hendra, and A. J. McQuillian, *Chem. Phys. Lett.* **26,** 163 (1974).

22. P. J. Hendra, I. D. M. Turner, E. J. Loader, and M. Stacey, *J. Phys. Chem.* **78,** 300 (1974).

23. T. A. Egerton, A. H. Hardin, Y. Kosiovovski, and N. Sheppard, *J. Catal.* **32,** 343 (1974).

24. T. A. Egerton, A. H. Hardin, and N. Sheppard, *Can. J. Chem.* **54,** 586 (1976).

25. R. P. Van Duyne, *J. Phys. (Paris) Colloq.* **52,** 239 (1977).

26. M. Fleischmann, P. J. Hendra, and A. J. McQuillian, *J. Chem. Soc. D,* 80 (1974).

27. R. L. Paul, A. J. McQuillan, P. J. Hendra, and M. Fleischmann, *J. Electroanal. Chem. Interfacial Electrochem.* **66,** 248 (1975).

28. N. Fleischmann, P. J. Hendra, A. J. McQuillan, R. L. Paul, and E. S. Reid, *J. Raman Spectrosc.* **4,** 269 (1976).

29. R. P. Cooney, P. J. Hendra, and M. Fleischmann, *J. Raman Spectrosc.* **6,** 264 (1977).

30. C. P. Cheng, J. D. Ludowise, and G. L. Schrader, *Appl. Spectrosc.* **34,** 146 (1980).

31. C. P. Cheng and G. L. Schrader, paper presented at 179th ACS National Meeting, Division of Colloid and Surface Chemistry, Houston, Texas, March 24–27 (1980).

32. H. Yamada and H. Naono, *Hyomen* **15,** 470 (1977).

33. H. Yamada, *Indian J. Pure Appl. Phys.* **16,** 159 (1978).

34. S. Efrima and H. Metiu, *J. Chem. Phys.* **70,** 1939 (1979).

35. Y. Yamamoto and H. Yamada, *J. Chem. Soc., Faraday Trans. 1* **74,** 1562 (1978).

36. T. Takenaka and T. Nakanaga, *J. Phys. Chem.* **80,** 475 (1976).

37. D. L. Jeanmaire, M. R. Suchanski, and R. P. Van Duyne, *J. Amer. Chem. Soc.* **97,** 1699 (1975).

38. M. R. Suchanski and R. P. Van Duyne, *J. Amer. Chem. Soc.* **98,** 250 (1976).

39. D. L. Jeanmaire and R. P. Van Duyne, *J. Amer. Chem. Soc.* **98,** 4029 (1976).

40. D. L. Jeanmaire and R. P. Van Duyne, *J. Amer. Chem. Soc.* **98,** 4034 (1976).

41. R. Dupeyrat, *Proc. 6th Int. Conf. Raman Spectrosc.,* Heyden, London, (1978).

42. P. J. Hendra, *The Spex Speaker* **19,** 1 (1974).

43. T. Takenaka and H. Fukuzaki, *J. Raman Spectrosc.* **8,** 51 (1979).

44. R. J. Thibeau, C. W. Brown, and R. H. Heidersbach, *Appl. Spectrosc.* **32,** 532 (1978).

45. W. L. Peticolas, E. W. Small, and B. Fanconi, in *Polymer Characterization Interdisciplinary Approaches,* C. D. Smith-Craver, Ed., Plenum, New York, 1971, p. 47.

46. M. G. Albrecht and J. A. Creighton, *J. Amer. Chem. Soc.* **99,** 5215 (1977).

47. B. Pettinger and U. Wenning, *Chem. Phys. Lett* **56,** 253 (1978).

48. For a review of the work up to 1978, see R. P. Van Duyne in *Chemical and Biological Applications of Lasers,* Vol. 4, C. B. Moore, Ed., Academic, New York, 1979, chap. 5.

49. D. L. Jeanmaire and R. P. Van Duyne, *J. Electroanal. Chem.,* **84,** 1 (1977).

50. C. Y. Chen, E. Burstein, and S. Lundquist, *Solid State Comm.* **32,** 63 (1979).

51. T. E. Furtak, *Solid State Comm.* **28,** 903 (1978).

52. A. Otto, *Surf. Sci.* **75,** 392 (1978).

53. B. Pettinger, V. Wenning, and C. M. Kolb, *Ber. Buns. Phys. Chem.* **82,** 1326 (1978).

54. J. A. Creighton, M. G. Albrecht, R. E. Hester, and J. A. D. Matthew, *Chem. Phys. Lett.* **55,** 55 (1978).

55. R. L. Birke, J. R. Lombardi, and J. I. Gersten, *Phys. Rev. Lett.* **43,** 71 (1979).

56. R. M. Hexter and M. G. Albrecht, *Spectrochim. Acta, Part A* **35,** 233 (1979).

57. E. Burstein, Y. J. Chen, C. Y. Chen, S. Lunquist, and E. Tosatti, *Solid State Comm.* **29,** 565 (1979).

58. J. C. Tsang, J. Kirtley, and J. A. Bradley, *Phys. Rev. Lett.* **43,** 727 (1979).

59. J. A. Creighton, C. G. Blatchford, and M. G. Albrecht, *J. Chem. Soc., Faraday Trans.* 2 **75,** 790 (1979).

60. E. Burstein and C. Y. Chen, *Proc. VIIth Int. Conf. Raman Spectrosc.,* W. F. Murphy, Ed., North Holland, Amsterdam, 1980, pp. 346–354.

61. M. Philpott, *J. Chem. Phys.* **62,** 1212 (1975).

62. F. W. King, R. P. Van Duyne, and G. C. Schatz, *J. Chem. Phys.* **69,** 4472 (1978).

63. S. Efrima and H. Metiu, *J. Chem. Phys.* **70,** 1602 (1979); *ibid,* p. 1939; *ibid,* p. 2297.

64. C.S. Allen and R. P. Van Duyne, *Chem. Phys. Lett.* **63,** 455 (1979).

65. M. Moskovits and D. P. Dilella, *Chem. Phys. Lett.* **73,** 500 (1980).

66. D. S. Wang, H. Chew, and M. Kerker, *Appl. Opt.* **19,** 2256 (1980).

67. M. Moskovits, *J. Chem. Phys.* **69,** 4159 (1978).

68. J. L. Callahan, R. K. Grasselli, E. C. Milberger, and H. A. Strecker, *Ind. Eng. Chem. Prod. Res. Dev.* **9,** 134 (1970).

69. B. C. Gates, J. R. Katzer, and G. C. A. Schuit, *Chemistry of Catalytic Processes,* McGraw-Hill, New York, 1979.

70. R. K. Grasselli and D. D. Suresh, *J. Catal.* **25,** 273 (1972).

71. A. W. Sleight, W. J. Linn, and K. Aykan, *Chem. Technol.* **8,** 235 (1978).

72. P. L. Villa, A. Szabo, F. Trifiro, and M. Carbuicchio, *J. Catal.* **47,** 122 (1977).

73. E. V. Hoefs, J. R. Monnier, and G. W. Keulks, *J. Catal.* **57,** 331 (1979).

74. J. G. Grasselli, M. A. S. Hazle, J. R. Mooney, and M. Mehicic, *Proc. 21st Colloq. Spectrosc. Int.,* in press.

75. E. Payen, J. Barbillat, J. Grimblot, and J. P. Bonnelle, *Spectrosc. Lett.* **11,** 997 (1978).

76. J. Medema, C. VanStam, V. H. J. DeBeer, A. J. A. Konings, and D. C. Koningsberger, *J. Catal.* **53,** 386 (1978).

77. F. R. Brown and L. E. Makovsky, *Appl. Spectrosc.* **31,** 44 (1977); F. R. Brown, L. E. Makovsky, and K. H. Rhee, *J. Catal.* **50,** 162 (1977); F. R. Brown, L. E. Makovsky, and K. H. Rhee, *J. Catal.* **50,** 385 (1977); F. R. Brown, L. E. Makovsky, and K. H. Rhee, *Appl. Spectrosc.* **31,** 563 (1977).

78. H. Knözinger and H. Jeziorowski, *J. Phys. Chem.* **82,** 2002 (1978).

79. F. P. J. M. Kerkhof, J. A. Moulyn, and R. Thomas, *J. Catal.* **56,** 279 (1979).

80. J. G. Grasselli, M. A. Hazle, and L. E. Wolfram, in *Molecular Spectroscopy,* A. West, Ed., Heyden, New York, 1977.

81. K. Aykan, *J. Catal.* **12,** 281 (1968); L. Y. Erman and E. L. Galperin, *Russ. J. Inorg. Chem.* (Engl. Transl.) **13,** 487 (1968).

82. V. Boudeville, F. Figueras, M. Forissier, J. L. Portefaix, and J. C. Vedrine, *J. Catal.* **58,** 52 (1979).

83. R. L. Farrow, P. L. Mattern, and A. S. Nagelberg, *Appl. Phys. Lett.* **36,** 212 (1980).

84. F. P. J. M. Kerkhof, J. A. Moulijn, and R. Thomas, *J. Catal.* **56,** 279 (1979).

Chapter Nine

Miscellaneous Raman Applications

9.1 RAMAN APPLICATIONS TO THE PETROLEUM INDUSTRY

The early application of Raman spectroscopy (pre-laser) was widespread in the petroleum industry. Ironically, laser excited Raman spectroscopy has not been utilized extensively in the petroleum field, mainly due to the sampling problems and fluorescence phenomena. There have been some applications, however, and the number is rapidly growing. The spectra of various petroleum fractions have been reported by Tooke (1). Low boiling products such as gasoline show many bands, mainly due to the presence of aromatic structures. Higher boiling fractions show broader, more poorly resolved bands.

Because of this complex and poorly resolved nature of the spectra, Raman spectroscopy has found greater application to the characterization of specific hydrocarbons than to the study of entire petroleum fractions. Although IR spectra of low molecular weight *n*-alkanes are very similar and tend to be dominated by CH stretching and bending vibrations, the Raman spectra show clearly the carbon skeleton vibrations and are, therefore, more useful for specific characterization of the molecules themselves. This is evident in Fig. 9.1, which shows the Raman spectra of the *n*-alkane series C_5 to C_8 (2).

The complexity and large number of bands in the "fingerprint region" of the spectra below 1300 cm^{-1} are caused by the presence of a number of different conformers. As the number of chain carbons increases in the homologous series, the number of conformers also increases, and the bands become broader and less well-resolved. In the solid state these bands are sharper, due to preferred conformations and crystallinity effects.

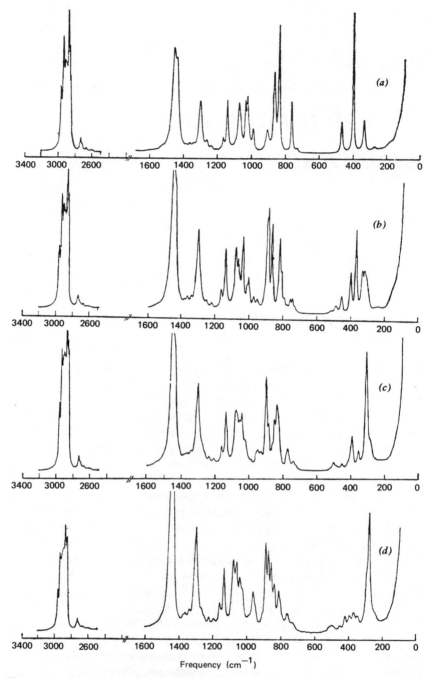

Figure 9.1 Raman spectra of *n*-alkanes, laser 514.5 nm: *(a)* *n*-pentane; *(b)* *n*-hexane; *(c)* *n*-heptane; *(d)* *n*-octane. (From ref. 2.)

148

Chain branching also causes large differences in the Raman spectra and allows easier skeletal elucidation than in the IR. Aromatic ring characterization has already been mentioned in Section 3.1. It was also pointed out that Raman spectroscopy is valuable in identifying groups such as —S—S, —C—S, and —O—O, which are found in petroleum product raw chemicals or intermediates.

In addition to the characterization of hydrocarbon types and functional groups, Raman spectroscopy has been applied to the examination of oil products, but there are many problems involved. For example, most oils are slightly or deeply colored. Some solid additives decompose in the beam, even at low laser power. Carbon tetrachloride dilution does not always halt these absorption effects, and they can sometimes cause local boiling. Fluorescence effects are frequently permanent—that is, they cannot be burned or quenched out—and prolonged irradiation usually results in sample destruction.

Many oil products and formulated additives contain colloidal particles, which, in a viscous liquid, cause Tyndall scattering and the appearance of plasma lines. In a mobile phase, they may cause spikes in the spectra produced by the Brownian motion of the particles as they move through the liquid. Filtration and/or centrifugation often are used to "clean up" this colloidal matter.

In spite of these problems, Raman spectroscopy has found application to oil characterization, for example, the differentiation of synthetic and natural lube oils (2).

Raman has also been used to "fingerprint" oil spills. In order to identify pollutant sources, oil spills have been analyzed using various techniques including IR (3–5), gas chromatography, fluorescence, and low temperature luminescence. Ahmadjian and Brown (6) have shown that Raman spectra can provide confirmation or additional evidence of an oil spill source. To eliminate the major sampling problem involved—that of the high fluorescent background—the oil samples were diluted with pentane, shaken with coconut charcoal, filtered, and the solvent was evaporated at ambient temperatures.

Fig. 9.2 shows the large differences in the spectra of a lubricating oil, a kerosene, and a No. 2 fuel oil. There obviously would be no difficulty in distinguishing among them. In the spectra of the three different No. 2 fuel oils shown in Fig. 9.3, however, the differences are far more subtle and occur mainly in the relative peak heights. It would be much more difficult to make a specific identification unless confirmatory evidence were present. In an actual oil spill sample, shown in Fig. 9.4 along with two suspect oils, the spectrum was a closer match with suspect A, and it was later confirmed that the spill was from the same source as suspect A. Thus Raman spectroscopy can be used as an auxiliary technique where a combination of fingerprinting methods are necessary to build stronger cases against suspected polluters. Modern computerized search techniques will be invaluable in this process.

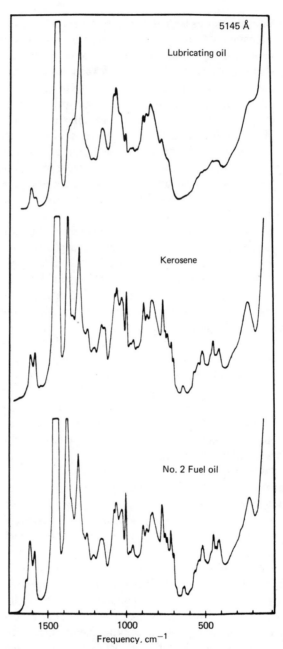

Figure 9.2 Raman spectra of a lubricating oil, a kerosene, and a No. 2 fuel oil. (From ref. 6.)

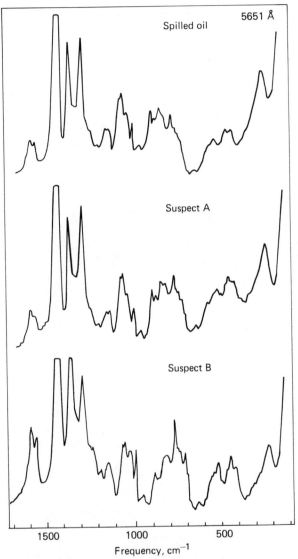

Figure 9.3 Raman spectra of three different No. 2 fuel oils. (From ref. 6.)

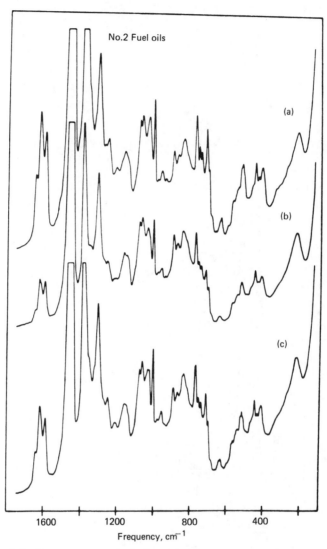

Figure 9.4 Raman spectra of a spilled oil and two suspect oils. (From ref. 6.)

In the petroleum industry, Raman spectra can also be valuable in characterizing and identifying additives, especially polymeric or sulfur-containing ones. Obremski (7) used the Raman technique in conjunction with IR to detect sulfur, determine how it was bonded into the system, and distinguish between three xanthates used as additives in high pressure lubricants.

Coates (2) has published the Raman spectra (as dialysis residues) of high molecular weight (MW = 1500) polyisobutene, polyalkyl methacrylate, and styrene-acrylate copolymers. He has also discussed the spectrum of Amoco 150, a sulfur-containing additive used as a copper passifier in industrial oils, which is shown in Fig. 9.5. C—S and S—S vibrations are evident at 635 cm^{-1} and ~ 500 cm^{-1}. The band at 731 cm^{-1} is probably associated with chain branching, most likely an isooctyl side chain.

Samples that are physically very small can be examined relatively easily by Raman spectroscopy because of the spatial resolution of the laser beam. Thus deposits on various automotive parts can be examined for identification and information on potential sources. Fig. 9.6 shows the spectrum, reported by Grasselli et al. (8), of a white material found in a carburetor from a car in a test fleet using experimental gasolines and motor oils. The carburetor was fouled with hydrocarbon deposit, but some tiny white crystals were also detected. These crystals carefully removed and examined directly with Raman spectroscopy by mounting them on a glass slide on a 180° viewing platform. The intense single peak at 998 cm^{-1} is characteristic of sodium sulfate. This identification could also have been made by either IR or x-ray techniques, but the ease and speed of the Raman analysis on this very small sample was a decided advantage.

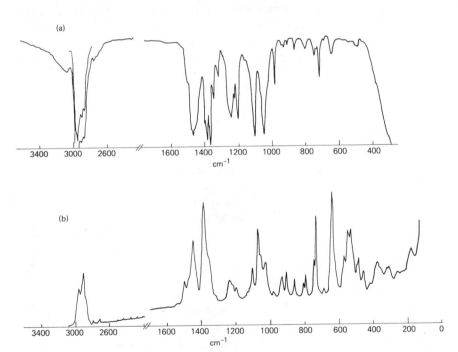

Figure 9.5 *(a)* Infrared spectrum of Amoco 150; *(b)* Raman spectrum of Amoco 150, laser 647.1 nm. (From ref. 2.)

153

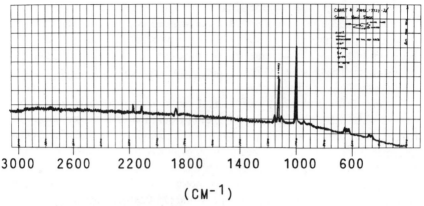

Figure 9.6 Raman spectrum of carburetor deposit. (From ref. 8.)

Raman spectroscopy has great potential for quantitative as well as qualitative analysis in the petroleum industry. It could be used, for example, to determine the amount of sulfurization, concentrations of materials in aqueous solutions or amounts of oil additives in formulations, isomer formations, hydrocarbon type determinations, or amounts of unsaturation. As instruments become less expensive and consequently more numerous within the petroleum industry, work will undoubtedly begin on these as well as other applications. The nonlinear Raman techniques (e.g., coherent anti-Stokes Raman scattering—CARS) may be of great assistance in overcoming the fluorescence problem.

9.2 BARBITURATES

Some common barbiturates have been examined by Raman spectroscopy to determine if they could be specifically identified. Analyses of barbiturates are important in clinical, forensic, and toxicological laboratories. As mentioned before, Raman has the advantage over other analytical techniques of sample-handling ease and ability to work in aqueous solutions. Willis et al. (9) studied eight of the more common barbiturates: phenobarbital, barbital, secobarbital, amobarbital, pentobarbital, butabarbital, mephobarbital, and hexobarbital. All barbiturates belong to the pyrimidine class and are used as the free form or the sodium salt of the acid.

Identifying the general class of barbiturates in mixtures containing other drugs, binders, and so on, is relatively easy, but distinguishing between the barbiturates is more difficult. A possible identification scheme is shown in Fig. 9.7. Position numbers refer to position on the pyrimidine ring.

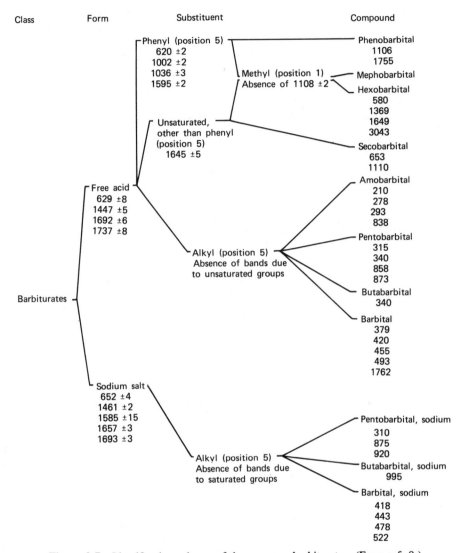

Figure 9.7 Identification scheme of the common barbiturates. (From ref. 9.)

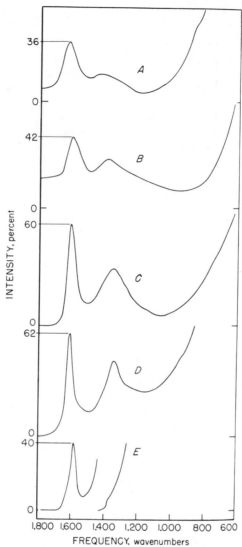

Figure 9.8 Raman spectra obtained on U.S. coals: *(a)* lignite, 74.2% carbon; *(b)* high volatile bituminous A, 83.1% carbon; *(c)* low volatile bituminous, 90.0% carbon; *(d)* anthracite, 94.2% carbon; *(e)* graphite, highly orientated pyrolytic. (From ref. 10.)

9.3 COALS

Although Raman spectra cannot usually be obtained with very darkly colored materials, there have been studies of coals (10, 11). The spectra of lignite, high volatile bituminous, low volatile bituminous, anthracite, and graphite (powders and polished solids) are presented in Fig. 9.8. The spectra are quite similar with bands at $\sim$ 1585 and $\sim$ 1360 cm^{-1}; the sharpness of the lines increases with increasing rank of the coals. These bands are attributed to graphitic or carbonized structures in the coal.

9.4 WATER POLLUTION

There has been some activity involving Raman spectroscopy in the area of water pollution. In theory the water samples can be examined directly; however, in practice, fluorescence often interferes. In these cases, organic materials can be either extracted directly into carbon tetrachloride or Freon or converted into an extractable species (12).

Ahmadjian and Brown (6) found that the addition of powdered charcoal to a dilute solution of petroleum oil removed the fluorescent compounds. In Section 9.1 we already discussed their work on the identification of petroleum products in oil spill situations. No. 2 fuel oils, kerosenes, lubricating oils, weathered oils, and actual spill oils were measured.

Cunningham et al. (13) investigated detection limits for various solutes in water. Solutions of nitrate, sulfate, carbonate, bicarbonate, monohydrogen phosphate, dihydrogen phosphate, acetate ion, and acetic acid were measured and the influence of experimental parameters on detection sensitivity determined. Laser beam intensity and solvent background intensity are the two most influential factors. Under ideal conditions the ionic and molecular species could be detected in the 5 to 50 ppm range.

Bradley and Frenzel (14) had previously reported the detection of benzene in water at concentrations of 50 ppm, and Baldwin and Brown (15) had found detection limits of 25 to 75 ppm for various inorganic species. Using remote sensing equipment, the lower limit of detection was even higher; Ahmadjian and Brown (16) could only detect 100 to 200 ppm of the inorganic species.

Ullman (17) investigated the Raman technique for water quality assessment and detected nitrite and sulfate at 20 ppm and 8.5 ppm, respectively. Various herbicides and plant growth regulators were also examined, but attempts to quantitate all of them were not successful.

To lower the minimal detectable concentration of pollutants, several workers took advantage of RRE, mentioned previously, which occurs when the exciting frequency approaches the frequency of an electronic absorption. This can increase the intensity of the solute spectrum without enhancing that of the water solvent. Brown and Lynch (18) showed the sensitivity of resonance Raman with identification of FD & C dyes in sodas and juice mixes at concentrations as low as 5 ppm. Van Haverbeke et al. (19) studied industrial fabric dyes and detected concentrations below 100 ppb and identified concentrations below 200 ppb. They simulated a "real world" situation by polluting river water with 288 ppb Superlitefast Rubine dye. Fig. 9.9*b* shows the spectrum between 1600 and 1000 cm^{-1}. For comparison, the spectrum of 24.3 ppm of the same dye in distilled water is shown in Fig. 9.9*a*.

Preresonance Raman was utilized by Thibeau et al. (20) to detect pesticides and fungicides in water. Preresonance Raman differs from resonance Raman in that the exciting frequency is situated within the band envelope, but not near the center. The pesticides and fungicides which were studied, all based on the nitrophenol structure, are listed in Table 9.1 along with observed Raman bands,

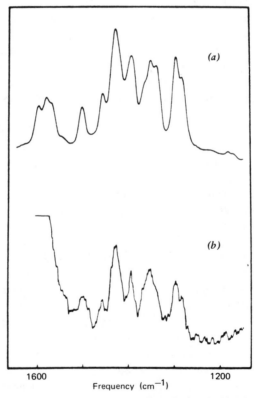

Figure 9.9 Resonance Raman spectra of: *(a)* Superlitefast Rubine in distilled water at 24.3 ppm; *(b)* Superlitefast Rubine in river water at 288 ppb. (From ref. 19.)

158

Table 9.1 Pesticides/Fungicides, Absorption Band Maxima, Minimum Detectable Concentrations, and Observed Raman Bands

Compound Name[a]	λ_a^b (nm)	E_a^c	E_0^d	Minimum Detectable Concentration (ppm)	Raman Bands Observed (cm^{-1})
2-Nitrophenol	414	1.43×10^3	7.85×10^2	0.8	824, 881, 1083, 1255, 1338
2, 4-Dinitrophenol	357	1.02×10^4	9.81×10^2	0.7	801, 964, 1318, 1350
2-Methyl-4, 6-dinitrophenol (DNOC) (Ditrosol)	366	7.29×10^3	1.43×10^3	0.9	1279, 1330
0, 0-Dimethyl-0-4-nitrophenyl phosphorothioate (methyl parathion)	278	2.86×10^3	3.74×10^1	7.0	1361
2, 6-Dichloro-4-nitroaniline (Dichloran)	364	1.19×10^4	2.42×10^2	0.4	1338
4, 6-Dinitro-2-sec-butylphenol (Dinoseb) (DNBP)	376	1.45×10^4	4.15×10^3	0.5	943, 1272, 1327

From ref. 20.

[a]Some trade names are given in parentheses.

[b]λ_a = position of absorption band maximum.

[c]E_a = molar absorptivity at band maximum.

[d]E_0 = molar absorptivity at excitation wavelength, 457.9 nm.

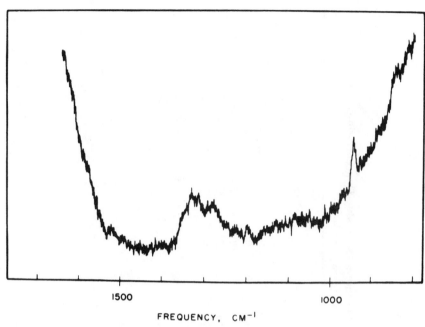

FREQUENCY, CM⁻¹

Figure 9.10 Raman spectrum of a 0.5 ppm solution of 4,6-dinitro-2-*sec*-butylphenol in 2×10^{-4} M NaOH, 457.9 nm excitation. (From ref. 20.)

absorption band maxima, and minimum detectable concentrations. The latter were defined as the smallest concentrations that gave strong bands with a signal-to-noise ratio of 3 or more. A typical example is the spectrum of a 0.5 ppm solution of 4,6-dinitro-2-*sec*-butylphenol, shown in Fig. 9.10. The spectrum is dominated by the 1640 cm⁻¹ band of water, but bands due to the compound itself are clearly evident at 943, 1272, and 1327 cm⁻¹ (symmetric NO_2 stretch). Measurement of lower concentrations of these samples should be possible with the use of excitation wavelengths closer to the absorption band maxima (approaching the resonance Raman condition). The future of resonance and preresonance studies such as these will depend on the development of inexpensive tunable laser radiation in the UV region, where a large number of interesting species are in resonance.

9.5   AIR POLLUTION

There has been considerable interest in the remote analysis of gaseous pollutants by Raman spectroscopy (21, 22). The current method projects the beam from a powerful laser into the atmosphere, then collects and analyzes the radiation back scattered along the line of sight with a conventional Raman spectrometer (23). One of the useful aspects of this Raman scattering method is that the distance of the pollutant from the detector can be related to the time delay between sending and receiving the returned pulse. This has led to the acronym for this technique: LIDAR, light detection and ranging. Initially equipment was very large (the unit was operated from a truck trailer) and expensive (24), but many improvements were made in second generation equipment (25). The schematic for a field tested remote Raman system is shown in Fig. 9.11.

Raman frequencies for some gaseous species of interest in pollution studies are shown in Table 9.2 (26). Detection limits vary widely among the species but, typically, 100 ppm of NO and SO_2 can be detected at a range of 300 m (23). There is considerable interest in trying to quantitate LIDAR measurements (27) as well as to increase sensitivities to <1 ppm. One possibility to greatly lower detection limits is the use of RRE (28) or of the rotational Raman effect, which in principle allow sensitivity to be increased by $\geq$ 10,000 over the vibrational Raman effect (29). LIDAR-type systems also have great potential for smokestack emission studies (30), and have even been used to characterize oil spills at sea (31).

9.6 FLAMES AND COMBUSTION

Before the early 1970s IR emission spectroscopy was one of the major spectroscopic tools for obtaining spectra of molecular species in flames, but it was

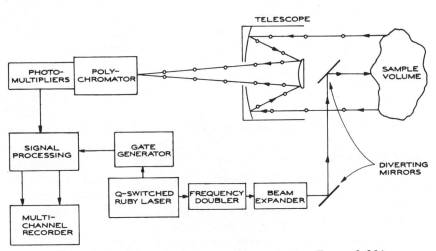

Figure 9.11 Schematic of a remote Raman system. (From ref. 26.)

Table 9.2   Raman Frequencies for Some Gaseous Species of Interest in Pollution Studies

Molecule	ν (cm^{-1})	Molecule	ν (cm^{-1})
CO_2	668	NO^+	2248
O_3	710	N_2	2273
SF_6	775	NO^+	2277
NH_3	950	N_2	2302
Aliphatics	987	NO^+	2305
Aromatics	992	N_2	2331
O_3	1043	CO	2349
SO_2	1151	H_2S	2611
CO_2	1242	O_3	2800
CO_2	1265	Aliphatics	2857
N_2O	1285	CH_4	2914
CO_2	1286	O_3	3050
NO_2	1320	Primary amines	3189
CO_2	1388	Primary amines	3256
CO_2	1409	NH_3	3331
CO_2	1430	Primary amines	3343
CO_2	1528	CO_2	3609
O_2	1556	H_2O	3652
O_3	1740	CO_2	3716
NO	1876	NO^+	4422
O_3	2105	N_2	4459
CO	2145	NO^+	4478
N_2^+	2175	N_2	4517
NO^+	2221	NO^+	4534
N_2O	2223	N_2	4575
N_2	2244	NO^+	4590
		N_2	4633

From ref. 26.

limited by the facts that homonuclear diatomic molecules do not give IR spectra, and that the temperature in the region above the flame is too low to allow the use of any but the most sensitive of the IR instruments. Raman spectroscopy overcomes many of these disadvantages: it does give spectra of N_2, O_2, and H_2; lack of sample brightness is a help; and it does not have H_2O and CO_2 spectra that mask large portions of the spectra. The principles and several applications of Raman studies of flames have been reviewed (32).

Arden et al. (33), using a Raman instrument and a flame isolation chamber, studied the combustion products of acetylene-oxygen, hydrogen-oxygen, ethane-oxygen, and propane-oxygen flames, and obtained spectra of CO_2, O_2, H_2O, and CO. In addition to the normal spectral lines, they found new bands appearing at higher temperature due to increased population of higher levels by thermal excitation. An example is the 1286 and 1388 cm^{-1} doublet of CO_2. At room temperature these two lines predominate, but as the temperature is raised there are fewer molecules in the ground state and the areas under these peaks get smaller; new bands first appear at $\Delta\nu = 1265$ and 1409 cm^{-1}, then at 1242 and 1430 cm^{-1}. For quantitative measurements the ratio of intensities of a pair of lines can be used once their relative population in any state has been determined as a function of temperature (i.e., a plot of line intensity ratio vs. temperature).

Flame temperature calculations have also been made by Stricker (34), who developed a method of high precision over a wide temperature range. A high degree of accuracy was also claimed by Schoenung and Mitchell (35), who compared Raman and thermocouple measurements, and by Beardmore et al. (36), who compared four methods of calculating the temperatures in a laminar gas/air flame from the Raman spectra for N_2.

The study of combustion processes by spontaneous Raman scattering has always been limited by intensity of the scattering species. It was thus natural to turn to nonlinear Raman techniques to overcome these problems. While, as mentioned in Chapter 11, for liquids or solids the dispersion causes problems of phase matching in CARS, for gases this is not the case. They are virtually dispersionless. Thus the phase matching conditions (Fig. 11.3) is met for a colinear geometry in gases. This geometry has certain problems for spectral resolution in flames, but these can be easily overcome by changing the beam geometry, (a technique known as BOXCARS (37). Coherent anti-Stokes Raman scattering provides good thermometry and species measurements, even in sooty flames (38, 39).

REFERENCES

1. P. B. Tooke, in *Infrared and Raman Spectroscopy*, Vol. 1, Part B, E. G. Brame and J. G. Grasselli, Eds., Dekker, New York, 1977.

2. J. P. Coates, in *Recent Analytical Developments in the Petroleum Industry*, D. R. Hodges, Ed., Applied Science Publishers, Essex, England, 1974.

3. C. W. Brown, P. F. Lynch, and M. Ahmadjian, *Appl. Spectrosc. Rev.* **9**, 223 (1975).

4. C. W. Brown, P. F. Lynch, M. Ahmadjian, and C. D. Baer, *Amer. Lab.* **7**, 59 (1975).

5. C. W. Brown and P. F. Lynch, *Anal. Chem.* **48**, 191 (1976).

6. M. Ahmadjian and C. W. Brown, *Anal. Chem.* **48**, 1257 (1976).

7. R. J. Obremski, *Spectra-Physics Raman Technical Bulletin,* Spectra-Physics, Mountain View, Calif., 1970.

8. J. G. Grasselli, M. A. S. Hazle, J. R. Mooney, and M. Mehicic, *Proc. 21st Colloq. Spectrosc. Int.,* Heyden, London, 1979.

9. J. N. Willis, R. B. Cook, and R. Jankow, *Anal. Chem.* **44,** 1228 (1972).

10. R. A. Friedel and G. L. Carlson, *Chem. Ind.* (London), 1128, (1971).

11. R. Tsu, H. J. Gonzalez, C. Hernandez, and C. A. Luengo, *Solid State Comm.* **24,** 809 (1977).

12. D. S. Lavery, in *Infrared and Raman Spectroscopy,* Vol. 1, Part B, E. G. Brame and J. G. Grasselli, Eds., Dekker, New York, 1977.

13. K. M. Cunningham, M. C. Goldberg, and E. R. Weiner, *Anal. Chem.* **49,** 75 (1977).

14. F. B. Bradley and C. A. Frenzel, *Water Res.* **4,** 125 (1970).

15. S. F. Baldwin and C. W. Brown, *Water Res.* **6,** 1601 (1972).

16. M. Ahmadjian and C. W. Brown, *Environ. Sci. Technol.* **7,** 452 (1973).

17. F. G. Ullman, U.S. National Technical Information Service, PB Report 261238, 1976.

18. C. W. Brown and P. F. Lynch, *J. Food Sci.* **41,** 1231 (1976).

19. L. Van Haverbeke, P. F. Lynch, and C. W. Brown, *Anal. Chem.* **50,** 315 (1978).

20. R. J. Thibeau, L. Van Haverbeke, and C. W. Brown, *Appl. Spectrosc.* **32,** 98 (1978).

21. C. K. N. Patel, *Science* **157,** 4364, (1978).

22. T. Haaka, *Denshi Gyutsu Sogo Kenkyusho Iho* **42,** 376 (1978).

23. H. A. Willis, in *Advances in Infrared and Raman Spectroscopy,* Vol. 2, R. J. H. Clark and R. Hester, Eds., Heyden, London, 1976, p. 132.

24. È. R. Schildkraut, *Amer. Lab.* **4,** 23 (1972).

25. T. Hirschfeld, E. R. Schildkraut, H. Tannenbaum, and D. Tannenbaum, *Appl. Phys. Lett.* **22,** 38 (1973).

26. S. K. Freeman, *Applications of Laser Raman Spectroscopy,* Wiley-Interscience, New York, 1974.

27. S. K. Poultney, M. L. Brumfeld, and J. H. Siviter, Jr., *Appl. Opt.* **16,** 3180 (1977).

28. O. Chamberlain, P. Robrish, and H. Rosen, U.S. National Technical Information Service, PB Report 262336, 1976.

29. W. H. Smith and R. A. King, U.S. National Technical Information Service, PB Report 273101, 1977.

30. K. Fredriksson, I. Lindgren, and S. Svanberg, *4th Joint Conf. Sens. Environ. Pollut. (Conf. Proc.),* 1977, published 1978, p. 446.

31. T. Sato, Y. Suzuki, H. Kashiwagi, M. Nango, and Y. Kakui, *Appl. Opt.* **17,** 3798 (1978).

32. M, Lapp and C. M. Penney, in *Advances in Infrared and Raman Spectroscopy,* Vol. 3, R. J. H. Clark and R. E. Hester, Eds., Heyden, London, 1977, p. 204.

33. W. M. Arden, T. B. Hirschfeld, S. M. Klainer, and W. A. Mueller, *Appl. Spectrosc.* **28,** 554 (1974).

34. W. Stricker, *Laser 77 (Seventy-Seven) Opto-Electron, Conf. Proc.*, 1977, p. 174.

35. S. M. Shoenung and B. E. Mitchell, U.S. National Technical Information Service, SAND Report 77-8722, 1977.

36. L. Beardmore, H. G. M. Edwards, D. A. Long, and T. K. Tan, *Lasers Chem., Proc. Conf.*, 1977, p. 79.

37. A. C. Eckbreth, *Appl. Phys. Lett.* **32,** 421 (1978).

38. I. R. Beattie, J. D. Black, and T. R. Gilson, *Combust. Flame* **33,** 101 (1978).

39. A. C. Eckbreth and R. J. Hall, *Combust. Flame* **36,** 87 (1979).

Chapter Ten

Raman Band Shapes as a Source of Information

Gordon (1) provided a formulation many years ago whereby information on orientational motion on the picosecond time scale could be obtained from the Fourier transform of Raman band shapes. An excellent summary of work up to 1974 may be found in the book edited by Lascombe (2).

10.1 BACKGROUND THEORY

In small molecules (approximately five atoms or fewer), reorientation provides the primary relaxation mechanism for allowed transitions. For larger molecules, vibrational relaxation mechanisms play an increasingly important role. The development of a coherent approach to such molecules has been the subject of several recent papers. The formulation most widely used is that of Nafie and Peticolas (3). They showed that in isotropic media the overall correlation function, $C(t)$, obtained from the Raman shape, may be conveniently separated into a reorientational part

$$Tr<\beta(t)\beta(0)>$$

where β is the anisotropic part of the polarizability, and

$$<Q(t)Q(0)>$$

a correlation function of all mechanisms of vibrational relaxation. This separation is achieved by computing the isotropic spectrum $I_{vib} = I_\parallel - \frac{4}{3}I_\perp$ and normalizing it to I_{vib}, whence

$$I_{vib} = (2\pi)^{-1} \int_{-\infty}^{\infty} <Q(0)Q(t)> \exp(i\omega t)\, dt$$

or, inversely,

$$<Q(0)Q(t)> = \int_{-\infty}^{\infty} I_{vib}(\omega) \exp(-i\omega t)\, d\omega$$

Similarly, the Fourier transform of the normalized $I_\perp$ (ω) band gives

$$<Tr[\beta(0)\beta(t)]><Q(0)Q(t)> = \int_{-\infty}^{\infty} I_\perp(\omega)\exp(-i\omega t)\, d\omega$$

Dividing the two results yields the reorientational correlation function, assuming that the reorientational and vibrational processes are uncorrelated.

Until recently, most workers have approached vibrational relaxation as a complication on Raman band shapes that needs to be removed if one is to study reorientational motion. Recent work, however, has shown that much valuable information about liquids might be derived from a study of vibrational relaxation processes by this method.

Before discussing specific results and interpretations, it is useful to define certain additional quantities connected with the band shape and correlation function. The second moment of a band M_2 or $<\omega^2>$ is obtained from

$$\int_{band} (\omega - \omega_0)^2 I(\omega)\, d\omega$$

where ω_0 is the frequency at the band center. One may also define a correlation time, τ, although this has been obtained in various ways (4). In general, τ should be obtained from

$$\int\limits_0^t C(t)\,dt$$

where the integration is carried out to the limit of reliability of the correlation function. For pure Lorentzian or pure Gaussian bands, there are other ways of obtaining τ.

Kubo (5) developed a general theory of relaxation processes that has been adapted to vibrational relaxation by several authors. Using this theory, one can show that correlation functions that involve "pure dephasing" processes are given by

$$C(t) = \exp\{-<\omega^2> [t\tau_c + \tau_c^2 \exp(\frac{-t}{\tau_c}) - 1]\}$$

This equation is often viewed in the short ($t << \tau_c$) and long ($t >> \tau_c$) modulation regimes or, more usefully, in what are known as the slow ($<\omega^2>^{1/2}\tau_c >> 1$) or fast ($<\omega^2>^{1/2}\tau_c << 1$) modulation regimes, whence it reduces to

$$C(t) = \exp(-1/2<\omega^2>t^2)$$

or

$$C(t) = \exp(-<\omega^2>\tau_c t)$$

respectively.

10.2 EXPERIMENTAL RESULTS

From the viewpoint of solutions of interest to the chemist, these dephasing mechanisms appear to predominate over energy relaxation (lifetime shortening) mechanisms (4). This is of value for the study of solution structure, as the change in environment of an excited molecule can be monitored in a rather unique way from a study of the correlation functions. As the modulation by the environment changes, both τ_c and $<\omega^2>$ are affected. This is known as "motional narrowing." If these effects could be studied systematically, they might lead to a probe of solution structures. Some examples of recent work are now given.

Yarwood et al. (4) have been examining acetonitrile, CH_3CN, in detail. Typical vibrational relaxation correlation functions for ν_1 of acetonitrile in carbon tetrachloride, denoted as $\phi_v(t)$, are shown in Fig. 10.1. One can see the initial curvature (slow modulation) at short times followed by a linear region (fast

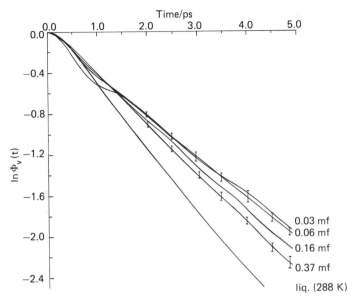

Figure 10.1 Comparison of ln $\phi_v(t)$ for ν_1 of acetonitrile at different concentrations. (From ref. 4.)

modulation). The motional narrowing, referred to above, can be seen here, leading to a slower decaying correlation function. This is shown more directly in Fig. 10.2, where values of τ_c and $<\omega^2>$ are shown as functions of concentration.

From data such as these one sees that there is a substantial "core" of intermolecular dephasing mechanism that remains even after long range interactions between solute molecules have been removed. This leads to the conclusion that short range repulsive interactions control relaxation for this mode, while dipole-dipole interactions are unimportant. Yarwood et al. (4) have shown that the relaxation for other modes of CH_3CN is quite different under the same conditions. For ν_3 (1375 cm^{-1}) τ_c increases on dilution in $CC\ell_4$. This points to dipole-dipole interactions as being important as a relaxation mechanism, as is resonant energy transfer.

Schroeder et al. (6) have carried out related studies on CH_3CN and CD_3CN, examining the pressure dependence of the ν_1 mode. Their results are also interpretable using the Kubo dephasing formalism. They have checked on the presence of resonance energy transfer by isotopic dilution measurements and agree that, for ν_1, this is not significant.

While most work on vibrational relaxation correlation functions has been for

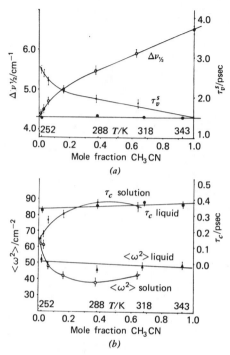

Figure 10.2 *(a)* Changes in $\Delta\bar{\nu}$ ½ and τ_ν^s for υ of acetonitrile as a function of temperature and concentration in carbon tetrachloride (solution data are at 288°K); *(b)* changes of $<\omega^2>$ and τ_c for ν_1 of acetonitrile (liquid and solution). (From ref. 4.)

small molecules, a few studies on larger molecules have indicated the potential for deriving information about condensed phases. Rothschild (7) has studied quinoline,

which should be large enough to have insignificant reorientation in the liquid phase during the first 5 psec. Fig. 10.3 shows vibrational correlation functions obtained from the 1033 cm^{-1} in phase bending mode in neat liquid (both IR and Raman) and 0.09 mole fraction in CS$_2$. A slight motional narrowing is seen in solution. In this case the IR and Raman correlation functions deviate after about 2.5 psec. While Rothschild does not discuss the reasons for this, a recent paper by Yarwood et al. (8) deals with this issue for acetonitrile. It appears that there may be a fundamental theoretical problem here but more experimental work is probably needed on this point. Rothschild's work on quinoline also includes a

170

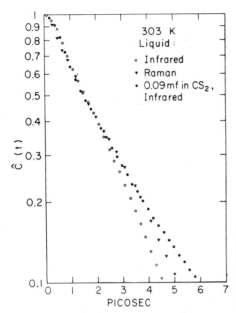

Figure 10.3 Vibrational correlation functions of the 1033 cm^{-1} in-plane bending (A') of quinoline at ambient temperature. Infrared: pure liquid and 0.09 mole fraction solution in CS$_2$, respectively. Raman: pure liquid. (From ref. 7.)

comparison of vibrational relaxation correlation functions for crystalline and glassy quinoline. The glassy state relaxes more rapidly, consistent with the idea that it offers the greatest number of distinct environments. There is potential here for studying glasses and phase transitions associated with them.

Bulkin and Brezinsky (9) have studied the CN stretching mode of 4-cyano-4'-octyloxybiphenyl (8OCB),

$$C_8H_{17}O \!\!-\!\!\bigodot\!\!-\!\!\bigodot\!\!-\!\! CN$$

which forms two liquid crystalline phases, two solid phases, as well as an isotropic liquid phase. The correlation functions obtained from the liquid crystalline phases are identical with those in the isotropic phase, indicating that the short range order in all of the fluid phases is similar. In CCℓ_4 or benzene solution, there is considerable motional narrowing, as seen in Fig. 10.4. Of further interest is the region from 0.7 to 1.5 psec, where the solution correlation functions differ from those of the pure liquid but there is no concentration dependence. This is the slow modulation regime. In CH$_3$SCN or CHCℓ_3 the effect of dilution is quite different (Fig. 10.5). No motional narrowing occurs. Clearly the solvation of

171

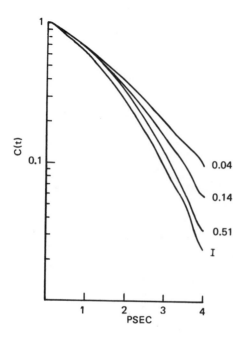

Figure 10.4 C_{vh} calculated from spectra of 8OCB dissolved in benzene in mole fractions of 0.51, 0.14, and 0.04. Also included is C_{vh} from spectra of the isotropic (I) phase. (From ref. 9.)

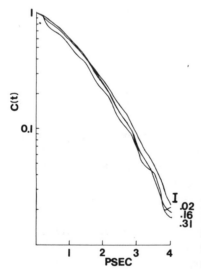

Figure 10.5 C_{vh} calculated from spectra of 8OCB dissolved in chloroform in mole fractions at 0.31, 0.16, and 0.02. Also included is C_{vh} from spectra of the isotropic (I) phase. (From ref. 9.)

the excited CN group is quite different in these solvents. These results indicate that there is a possibility of using vibrational relaxation correlation functions as a probe of solvent-solute interactions.

Band shape analysis and resolution studies have also been useful in obtaining information on conformational structures of polymers. Maddams and co-workers (10, 11) applied such techniques to obtain information on the configuration and conformation of poly(vinylchloride). In a separate study (12) they also measured the profiles of eight well-defined Raman bands to establish that Raman band shapes are a reasonable approximation to the commonly assumed Lorentzian form and, therefore, that curve fitting procedures to separate overlapping peaks are feasible.

REFERENCES

1. R. G. Gordon, *J. Chem. Phys.* **43**, 1307 (1965).

2. J. Lascombe, Ed., *Molecular Motions in Liquids*, Reidel, Dordrecht, Netherlands, 1974.

3. L. A. Nafie and W. Peticolas, *J. Chem. Phys.* **57**, 3145 (1972).

4. J. Yarwood, R. Arndt, and G. Döge, *Chem. Phys.* **25**, 387 (1977).

5. R. Kubo, in *Fluctuations, Relaxation, and Resonance in Magnetic Systems*, D. Ter Haar, Ed., Oliver and Boyd, Edinburgh, 1962.

6. J. Schroeder, D. H. Schiemann, P. T. Sharko, and J. Jones, *J. Chem. Phys.* **68**, 3215 (1977).

7. W. G. Rothschild, *J. Chem. Phys.* **65**, 455 (1976).

8. J. Yarwood, P. L. James, G. Döge, and R. Arndt, *Faraday Discuss. Chem. Soc.* **66**, 252 (1979).

9. B. J. Bulkin and K. Brezinsky, *J. Chem. Phys.* **69**, 15 (1978).

10. A. Baruya, A. D. Booth, W. F. Maddams, J. G. Grasselli, and M. A. S. Hazle, *J. Polym. Sci., Polym. Lett. Ed.* **14**, 329 (1976).

11. M. E. R. Robinson, D. I. Bower, and W. F. Maddams, *Polymer* **17**, 355 (1976).

12. C. Baker, W. F. Maddams, J. G. Grasselli, and M. A. S. Hazle, *Spectrochim. Acta, Part A* **34**, 761 (1978).

Chapter Eleven

Nonlinear Effects

With the normal Raman effect, there is a linear dependence of polarization on field strength; that is, as the laser power is increased, the Raman signal is proportionally increased. At sufficiently high laser power this need not be the case.

The polarization can be written as a (somewhat simplified) expansion in the field strength

$$P = \chi^{(1)} E + \chi^{(2)} E^2 + \chi^{(3)} E^3$$

The susceptibilities χ^i fall off rapidly, typically being reduced by 10^{10} in each successive term. Normally then, only the $\chi^{(1)}$ (sometimes called α) term is important. When E exceeds approximately 10^9 Vm^{-1}, however, the nonlinear contributions can be expected. Such fields are available from many laser systems. Indeed, the nonlinear effects were observed in the early days of laser excited Raman spectroscopy. Only recently have they been applied to chemical problems.

A number of nonlinear effects have been discovered. These are summarized in Table 11.1 (1). Included in the table is normal or spontaneous Raman scattering, as well as stimulated Raman scattering.

11.1 STIMULATED RAMAN SCATTERING

Stimulated Raman scattering has great utility in producing sources of a particular frequency (2). It may also be an effective method to study vibrational lifetimes (3, 4). While conversion efficiency is high, only certain lines, usually the most intense, will emit in the stimulated mode. This is because, after one reaches the threshold power for stimulated emission of one mode, all further increases in

Table 11.1 Nonlinear Effects

Effect	Raman Efficiency	Discrimination Against Fluorescence	Trace Analysis Capability	Determination of Low Number Densities	Comments
Normal Raman[a]	Low	Low	Moderate–low	Moderate–low (100 mtorr)	General application
Stimulated Raman	High	High	Very low	Very low (20 atm)	Limited usefulness
Inverse Raman	High	High	Low	Low (? 10 torr)	Good for condensed phase, not sensitive for gas analysis at low pressure
Hyper-Raman	Very low	High	Very low	Very low (~ 1 atm)	Sensitivity very low, good for some structural anal
Raman-induced Kerr effect	High	High	Moderate (?)	Moderate (?)	Too early for good assessment but has definite potential
CARS[a]	High	Very high	Low (at present[b])	High (< 1 mtorr)	Excellent for high resolution and analysis of gases at low pressure, problems with trace analysis (low ppm)

From ref. 1.

[a]These assessments do not include improvements effected by electronic resonance enhancements.

[b]The limit to CARS for trace analysis is the interference from background generation of solvent or diluent gas. At this writing, there are a few ideas, not yet fully tested, that may markedly reduce or eliminate background generation.

175

power go into that mode. There has been an attempt to take advantage of this to elucidate the structure of water (5).

11.2 INVERSE RAMAN AND STIMULATED RAMAN GAIN SPECTROSCOPY

The inverse Raman effect (6) occurs as follows. If a sample is simultaneously irradiated with an intense laser beam of frequency v_0, and a continuum from v_0 to $v_0 + 4000$ cm^{-1} (the typical range of a vibrational spectrum), absorption is observed at frequencies v_i characteristic of the Raman active modes. Emission is also observed at v_0. This is an inverse effect. In principle, it would have an advantage over normal Raman spectroscopy in that fluorescence would not be important. In practice, there have been few applications. This is because the absorption is weak and it is somewhat difficult to get the laser pulse and continuum to coincide spatially and temporally. A number of approaches (7, 8) to solve this problem have been demonstrated; they have lowered detection limits considerably.

There has been some recent work on stimulated Raman spectroscopy using low power CW lasers, as well as using lower average power pulsed lasers, which does show promise for chemical applications. This is primarily the work of Owyoung (9, 10) on a technique known as stimulated Raman gain spectroscopy. As shown below, this can also be done experimentally as inverse Raman spectroscopy.

Fig. 11.1 shows schematically the original Owyoung experiment using a CW He—Ne laser and a CW dye laser. The dye laser is chopped and tuned through the region of vibrational frequency differences from the He—Ne laser. The two beams are brought together by a beamsplitter in a temporal and spatial coincidence in the sample. One can measure either gain or loss as a function of frequency of the exiting beam. Because this beam is coherent, spatial filtering can be used to remove fluorescence.

Nestor (11) has done the same experiment using a pulsed nitrogen laser and two dye lasers. In this case a differential amplifier is used, with one input being the exit probe beam pulse, the other a split-off signal from the pulse before it enters the sample.

These techniques have thus far only been used on test samples. However, for gas phase spectra they look very promising. In liquids there may be some cases where the significant improvement in resolution obtainable from the tunable laser versus a monochromator is useful in solving a problem. There is interest in applying these methods to resonance Raman problems, but thus far problems of thermal blooming have not been solved.

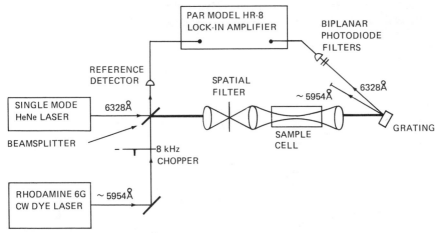

Figure 11.1 Schematic diagram of the CW stimulated Raman gain spectroscopy experiment. (From ref. 10.)

11.3 HYPER-RAMAN EFFECT

The hyper-Raman effect(a $\chi^{(2)}$ effect) gives rise to Raman scattering at frequencies of $\nu_0 \pm \nu_i$. It is of interest because selection rules for hyper-Raman scattering may be different from those of either normal Raman or IR spectroscopy. An example is shown in Table 11.2 (11). Vogt and Neumann applied this to observe Raman inactive modes of cubic crystals (12). The effect is extremely weak and the instrumentation complex. A computer controlled spectrometer has been described (13).

11.4 COHERENT ANTI-STOKES RAMAN SCATTERING

Coherent anti-Stokes Raman scattering (CARS) is a $\chi^{(3)}$ effect. It has received more attention than any of the other nonlinear effects (14). The bulk of the work thus far has been of a very exploratory nature, and the potential for solving problems is still a matter of some controversy. CARS generates coherent photons, and the efficiency for the CARS process can be significantly higher than for spontaneous Raman scattering.

Coherent anti-Stokes Raman scattering may be done with either pulsed or CW lasers, but most work to date has been done with pulsed lasers. If two beams of frequencies ω_ℓ and ω_s are focused in a sample at sufficient power, a coherent

Table 11.2   Infrared, Raman, and Hyper-Raman Activity of Fundamental Vibrations in Sulphur Hexafluoride (SF_6)

Symmetry Species	x, y, z	α	Components of β	Number of Distinct Frequencies
A_{1g}		$\alpha_{xx} + \alpha_{yy} + \alpha_{zz}$		1
A_{2g}				0
E_g		$(\alpha_{xx} + \alpha_{yy} - 2\alpha_{zz},$ $\alpha_{xx} - \alpha_{yy})$		1
F_{1g}				0
F_{2g}		$(\alpha_{xy}, \alpha_{xz}, \alpha_{yz})$		1
A_{1u}				0
A_{2u}			β_{xyz}	0
E_u				0
F_{1u}	(x, y, z)		$(\beta_{xxx}, \beta_{yyy}, \beta_{zzz}),$ $(\beta_{xyy} + \beta_{zzx}, \beta_{yzz}$ $+ \beta_{xxy}, \beta_{zxx} +$ $\beta_{yyz})$	2
F_{2u}			$(\beta_{xyy} - \beta_{zzx}, \beta_{yzz}$ $- \beta_{xxy}, \beta_{zxx} -$ $\beta_{yyz})$	1

Taken in part from a table in D. Long, *Raman Spectroscopy*, McGraw-Hill, New York, 1976.

beam of frequency $\omega_{as} = 2\omega_\ell - \omega_s$ may be generated in the medium. Many factors determine whether this conversion to ω_{as} takes place. These include the presence of molecular resonances at frequency $\omega_\ell - \omega_s$, and the properties of these resonances.

It is possible to derive an equation for the efficiency ε of the CARS process, and it is instructive to examine the terms in this expression. The efficiency is proportional to

$$L^2 \left[\frac{\sin (\Delta kL/2)}{\Delta kL/2} \right]^2$$

where L is the length over which the beams are mixed through the sample and Δk is the mismatch between the momentum vectors such that

$$\Delta k = 2k_\ell - k_s - k_{as}$$

Fig. 11.2 shows how ε varies with Δk at constant L, indicating that it is maximized at $\Delta k = 0$. This is known as the phase matching condition, and it imposes a very strict experimental condition on CARS. To achieve $\Delta k = 0$, the beams must be crossed at angle θ (Fig. 11.3). For gases this condition is not severe, but for liquids the angle tolerance for CARS is ≤1°. ε is proportional to $(\chi^{(3)})^2$. As mentioned above, the conversion efficiency depends on the occurrence of resonance at $\omega_\ell - \omega_s$; however, it can occur even in the absence of such resonances. One way of expressing this is to separate $\chi^{(3)}$ into resonant and nonresonant parts:

$$\chi^{(3)} = \chi^R + \chi^{NR}$$

It is χ^R that is related to Raman spectroscopic transitions, but χ^{NR} may provide an intense background, limiting the possibility of detecting χ^R. The third-order

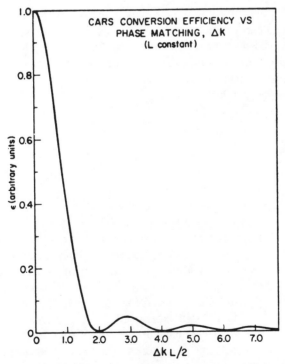

Figure 11.2 Coherent anti-Stokes Raman scattering conversion efficiency versus phase matching (Δk) for constant interaction length. (From ref. 16.)

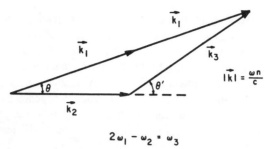

$$2\omega_1 - \omega_2 = \omega_3$$

Figure 11.3 Wave vector diagram for phase matching ($\Delta k = 0$). (From ref. 16.)

nonlinear susceptibility has another important property, namely, χ^R has both real and imaginary parts, that is,

$$\chi^{(3)} = \chi' + i\chi'' + \chi^{NR}$$

The imaginary part of $\chi^{(3)}$ is associated with the normal Raman transition probability, the real part with the nonlinear refractive index. These have usual forms shown in Fig. 11.4. If these were the only factors CARS band shapes would look much like normal Raman band shapes. In the presence of a significant nonresonant background, this can change.

This is simply seen by squaring $\chi^{(3)}$.

$$(\chi^{(3)})^2 = (\chi' + \chi^{NR})^2 + (\chi'')^2$$
$$= (\chi')^2 + 2\chi'\chi^{NR} + (\chi^{NR})^2 + (\chi'')^2$$

The mixing term $\chi'\chi^{NR}$ can distort normal band shapes leading to minima in the CARS spectra. An approach to suppressing the nonresonant background by polarization techniques has been described (15).

Tolles and Turner (16) have assessed the performance capabilities of CARS versus spontaneous Raman and absorption spectroscopy for the analysis of gases. They find that CARS offers advantages under certain conditions of temperature and pressure, particularly for a major component of a mixture below 1 atm total pressure. The calculations they describe are quite useful to an understanding of signal-to-noise ratio in a CARS experiment.

Rogers and co-workers (17) have developed a CARS spectrometer for application to analytical problems of solutions. They have concentrated their efforts on situations in which resonance enhanced CARS signals are observable. The range of solutions for which this is the case is increased by their use of UV

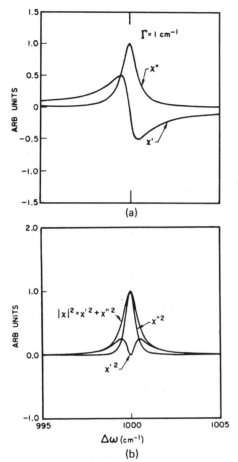

(a)

(b)

Figure 11.4 *(a)* Plot of the real (χ') and imaginary (χ'') parts of the third-order susceptibility for a resonance centered at 1000 cm^{-1} Raman shift and a line width of 1.0 cm^{-1} (note that χ' is negative in one-half of the frequency domain); *(b)* square of the quantitites plotted in *(a)*; χ^2 is directly proportional to the CARS signal (note that at the center of the resonance the real part of the susceptibility vanishes). (From ref. 16.)

excitation. To facilitate the operation of the spectrometer, expecially in regard to the phase matching angle condition, a number of computer controlled stepping motors are utilized. These include motors to move optics as necessary for phase matching. The computer "learns" the conditions necessary for each solvent and sets these automatically.

These authors have used the micro flow cell technique in resonance enhanced CARS to good advantage. This minimizes buildup of photo-decomposed products, thermal gradients, while permitting the study of excited molecules or intermediates. One possibility for application of the microflow cell in this spectrometer is as a liquid chromatograph detector. Further work in this area is needed.

181

Dutta et al. (18) have carried out resonance CARS experiments on a biological system, flavin adenine dinucleotide, where the electronic resonance is with a flavin absorption band. This is a highly fluorescent system, and the CARS data make a convincing case for the ability of CARS to obtain good Raman spectra despite this. Resonance enhanced CARS has been discussed and briefly reviewed by Morris and Wallan (19).

REFERENCES

1. A. Harvey, J. R. McDonald, and W. M. Tolles, *Prog. Anal. Chem.* **8**, 211 (1976).
2. M. J. Colles and G. E. Walrafen, *Appl. Spectrosc.* **30**, 463 (1976).
3. A. Laubereau, D. Vander Linde, and W. Kaiser, *Phys. Rev. Lett.* **28**, 1162 (1972).
4. W. T. Barnes and F. E. Lytle, *Appl. Phys. Lett.* **34**, 509 (1979).
5. G. E. Walrafen, *J. Chem. Phys.* **64**, 2699 (1976).
6. W. J. Jones and B. P. Stoicheff, *Phys. Rev. Lett.* **13**, 657 (1964).
7. W. Werncke, J. Klein, A. Lau, K. Lenz, and G. Hunsalz, *Opt. Comm.* **11**, 159 (1974).
8. J. P. Haunlter, G. P. Ritz, D. G. Wallan, K. Dien, and M. D. Morris, *Appl. Spectrosc.* **34**, 144 (1980).
9. A. Owyoung, *IEEE J. Quantum Electron.* **14**, 192 (1978).
10. A. Owyoung and E. D. Jones, *Opt. Lett.* **1**, 152 (1977).
11. J. R. Nestor, *J. Chem. Phys.* **69**, 1778 (1978).
12. H. Vogt and G. Neumann, *Phys. Stat. Sol.* **92**, 57 (1979).
13. M. G. French and D. A. Lanz, *J. Raman Spectrosc.* **3**, 391 (1975).
14. W. M. Tolles, J. W. Nibler, J. R. McDonald, and A. B. Harvey, *Appl. Spectrosc.* **31**, 253 (1977).
15. J. J. Song, G. L. Eesely, and M. D. Levenson, *Appl. Phys. Lett.* **29**, 567 (1976).
16. W. M. Tolles and R. D. Turner, *Appl. Spectrosc.,* **31**, 96 (1977).
17. L. B. Rogers, J. D. Stuart, L. P. Goss, T. B. Malloy, Jr., and L. A. Carreira, *Anal. Chem.* **49**, 960 (1977).
18. P. K. Dutta, J. R. Nestor and T. G. Spiro, *Proc. Nat. Acad. Sci. US* **74**, 4146 (1977).
19. M. D. Morris and D. J. Wallan, *Anal. Chem.* **51**, 182A (1979).

Chapter Twelve

Where Have We Come From, Where Are We Going?

The Raman effect was discovered at a propitious time in the history of chemistry. As the quantum theory of molecules developed, chemists were increasingly anxious for molecular structure information. In many cases, theory did not provide a basis for choice between alternative structures—linear versus bent, planar versus nonplanar, and so on. Using the link between group theory, symmetry, and spectroscopy, Raman spectroscopy, often in partnership with IR spectroscopy, provided many answers. Up to about 1970, such research dominated the application of Raman spectroscopy in chemistry.

The 1970s have been a period of change. As we have seen in the foregoing chapters, there have been many applications of the Raman effect to complex molecular systems. In part this has happened because other spectroscopic and diffraction techniques were providing more straightforward answers to the simpler problems. But in part it results as well from the maturity of Raman as a technique. The large volume of data and expertise on small molecules has allowed interpretable data to be obtained for biological systems, polymers, ordered condensed phases, and multicomponent systems.

The 1970s have also witnessed many developments in instrumentation, the fruits of which will be most apparent in the future. One for which the impact is already clear is computer support for data acquisition, reduction, and interpretation. We have also been able to use the developments in molecular orbital calculations on large molecules as aids in data interpretation.

Toward the end of the last decade, we saw the emergence of Raman spectroscopy as a technique for studying dynamics. This comes on many time scales. From Raman band shapes we glean information about picosecond processes. Using vidicons and pulsed lasers, nanoseconds to millisecond processes can be examined. And using rapid scanning techniques, we can probe reactions in the 0.1 to 10 sec range. As there are relatively few good techniques for obtaining dynamic information in condensed phases with the structural detail offered by Raman spectroscopy, we can expect to see continued development along these lines.

Thus in the 1980s we may expect to recover the multiplex advantage in Raman spectroscopy that was given up for convenience when photographic plates were replaced by photomultiplier tubes. Some laboratories already use detector arrays, as described in Chapter 2, but these can be expected to become more sophisticated, capable of higher resolution, and lower in price.

Computer based analysis of data, mainly through improved library search systems algorithms, is proving to be important at the transition from the 1970s to the 1980s in IR spectroscopy. Such techniques will naturally come to Raman systems soon. We may also see, in the 1980s, some increased use of theory for interpretation of relative intensities of bands on a routine basis.

New sources for Raman spectroscopy will surely emerge in the next decade and have great impact on the field. The efforts are now underway to develop a variety of improved UV and vacuum UV CW lasers; these will result in the capability to carry out resonance Raman experiments on virtually any material in a routine fashion. When combined with gating techniques to suppress fluorescence, this will greatly enhance the sensitivity of Raman spectroscopy at lower concentrations. It seems possible that with sophisticated computational backup, one may also be able to do intensity corrections on such resonance enhanced data to provide quantitative information on concentrations.

In addition to laser sources, we may expect to see some use of the storage ring sources in the vacuum ultraviolet, now either in operation or under construction, for resonance Raman spectroscopy in the hard vacuum UV regions. In addition to intensity, continuous tunability, and polarization, these sources have potential (not yet fully developed) for sophisticated time resolved experiments on the nanosecond time scale. The negative side of this, however, comes in the considerable expense in carrying out such experiments. It is not clear yet whether such an expense will be justified.

We have treated the nonlinear Raman effects briefly in this book. To date they have found few applications in chemistry. This will likely change in the 1980s. The simplicity of instrumentation, the high intensity, and the discrimination against fluorescence offered by these techniques make it likely that they may even come to dominate spontaneous Raman spectroscopy in liquid and gas phase problems. Their application in solids is much less clear and surely warrants further work.

184

These developments are all experimental. What of theory? Some of the most sophisticated quantum chemistry now underway deals with condensed phase theory. We can hope that this will result in the ability to more quantitatively interpret effects seen in condensed phase spectra—changes in band intensities, frequencies, and shapes—under varying conditions of solvent, temperature, pressure, viscosity, and such.

For some of this, we will need the results of other theoretical and experimental work now underway on the dynamics of intermode intramolecular energy transfer. Particularly in large molecules, this may be a factor in interpreting subtle effects in their Raman spectra.

There has been some access to new vibrational levels through new selection rules as the nonlinear techniques have evolved and as circularly polarized radiation has been used. We may hope that theoretical work in this area of new selection rules also evolves in the future.

Finally, we may hope that the next decade will close with the result of these theoretical and experimental advances providing a more complete network of links between chemistry and spectroscopy.

Index

This index relates to the main text and does not cover the references at the end of each chapter. The numbers of the pages of which spectra are reproduced are underlined. Readers will find the detailed Table of Contents at the front of the book valuable as a suipplement to this index.

197